385-1
52

PURE AND APPLIED MATHEMATICS

A Program of Monographs, Textbooks, and Lecture Notes

LECTURE NOTES
IN PURE AND APPLIED MATHEMATICS

1. *N. Jacobson,* Exceptional Lie Algebras
2. *L.-Å. Lindahl and F. Poulsen,* Thin Sets in Harmonic Analysis
3. *I. Satake,* Classification Theory of Semi-Simple Algebraic Groups
4. *F. Hirzebruch, W. D. Newmann, and S. S. Koh,* Differentiable Manifolds and Quadratic Forms
5. *I. Chavel,* Riemannian Symmetric Spaces of Rank One
6. *R. B. Burckel,* Characterization of C(X) Among Its Subalgebras
7. *B. R. McDonald, A. R. Magid, and K. C. Smith,* Ring Theory: Proceedings of the Oklahoma Conference
8. *Yum-Tong Siu,* Techniques of Extension of Analytic Objects
9. *S. R. Caradus, W. E. Pfaffenberger, and Bertram Yood,* Calkin Algebras and Algebras of Operators on Banach Spaces
10. *Emilio O. Roxin, Pan-Tai Liu, and Robert L. Sternberg,* Differential Games and Control Theory
11. *Morris Orzech and Charles Small,* The Brauer Group of Commutative Rings
12. *S. Thomeier,* Topology and Its Applications
13. *Jorge M. López and Kenneth A. Ross,* Sidon Sets
14. *W. W. Comfort and S. Negrepontis,* Continuous Pseudometrics
15. *Kelly McKennon and Jack M. Robertson,* Locally Convex Spaces
16. *M. Carmeli and S. Malin,* Representations of the Rotation and Lorentz Groups: An Introduction
17. *George B. Seligman,* Rational Methods in Lie Algebras
18. *Djairo Guedes de Figueiredo,* Functional Analysis: Proceedings of the Brazilian Mathematical Society Symposium
19. *Lamberto Cesari, Rangachary Kannan, and Jerry D. Schuur,* Nonlinear Functional Analysis of Differential Equations: Proceedings of the Michigan State University Conference
20. *Juan Jorge Schäffer,* Geometry of Spheres in Normed Spaces
21. *Kentaro Yano and Masahiro Kon,* Anti-Invariant Submanifolds
22. *Wolmer V. Vasconcelos,* The Rings of Dimension Two
23. *Richard E. Chandler,* Hausdorff Compactifications
24. *Stanley P. Franklin and Barbara V. Smith Thomas,* Topology: Proceedings of the Memphis State University Conference
25. *S. K. Jain,* Ring Theory: Proceedings of the Ohio University Conference

Other volumes in preparation

RING THEORY

Proceedings of the Ohio University Conference
May 1976

Edited by

S. K. JAIN
Professor of Mathematics
Ohio University
Athens, Ohio

With the cooperation of

KLAUS E. ELDRIDGE
Associate Professor of Mathematics
Ohio University
Athens, Ohio

MARCEL DEKKER, INC. New York and Basel

MARCEL DEKKER, INC.

270 Madison Avenue, New York, New York 10016

LIBRARY OF CONGRESS CATALOG CARD NUMBER: 76-56701

ISBN: 0-8247-6577-X

Current printing (last digit):
10 9 8 7 6 5 4 3 2 1

PRINTED IN THE UNITED STATES OF AMERICA

PREFACE

In January 1976 the Institute for Medicine and Mathematics at Ohio University gave us an initial grant to organize a conference in ring theory. With the additional financial support from the Ohio University Research Institute and the Mathematics Department, the conference was held on May 13 and 14, 1976. This volume contains the "fruits" of the conference, primarily representing the ring theory activity in the region - region being loosely defined as within 600 miles of Athens. Papers of some of the invited speakers who could not attend are also included.

We thank the participants, contributors, and referees for prompt responses which allowed us to maintain a tight schedule, thus making the conference such a success. Our thanks are due also to Professor Norman Cohn, Director, Ohio University Research Institute, and Professor R.G. Helsel (former Chairman of the Department of Mathematics) for their financial assistance and co-sponsorship of the conference, and we owe a note of thanks to our colleagues and to some of the senior graduate students who assisted us with the many details in organizing the conference. We also thank Marcel Dekker for undertaking the publishing of the volume and Mrs. Stephanie Goldsberry for the excellent typing. However, our biggest thanks go to Professor John M. Worrell, Jr., Director of the Institute for Medicine and Mathematics, and Dr. G.M. Reed, Associate Director, whose initial suggestions and encouragement made the conference possible.

Ohio University
Athens, Ohio
October, 1976

S. K. Jain
and
Klaus E. Eldridge

PARTICIPANTS

S. A. AMITSUR, Hebrew University, Jerusalem, Israel

CHRISTINE AYOUB, Pennsylvania State University, University Park, Pennsylvania

ROBERT L. BLAIR, Ohio University, Athens, Ohio

WILLIAM BLAIR, Northern Illinois University, Dekalb, Illinois

TOM CLARK, Kent State University, Kent, Ohio

JOHN COZZENS, Rider College, Trenton, New Jersey

DAVID CUSICK, Marshall University, Huntington, West Virginia

R. KEITH DENNIS, Cornell University, Ithaca, New York

KLAUS E. ELDRIDGE, Ohio University, Athens, Ohio

D. FIELDHOUSE, University of Guelph, Guelph, Ontario, Canada

JOE W. FISHER, University of Cincinnati, Cincinnati, Ohio

EDWARD FORMANEK, University of Chicago, Chicago, Illinois

SURESH C. GOEL, Ohio University, Athens, Ohio

V. K. GOEL, Ohio University, Athens, Ohio

WILLIAM G. GUSTAFSON, Indiana University, Bloomington, Indiana

NIKI HASA, Kent State University, Kent, Ohio

IRVIN ROY HENTZEL, Iowa State University, Ames, Iowa

S. HSIA, Ohio University, Athens, Ohio

S. K. JAIN, Ohio University, Athens, Ohio

MARCIA JONES, Kent State University, Kent, Ohio

MICHAEL DAVID KONRAD, Ohio University, Athens, Ohio

GÜNTER KRAUSE, University of Manitoba, Winnipeg, Canada

EDWARD KWAK, Ohio University, Athens, Ohio

JOHN LAWRENCE, University of Chicago, Chicago, Illinois

WALLACE S. MARTINDALE, 3rd, University of Massachusetts, Amherst, Massachusetts

DONALD NORRIS, Ohio University, Athens, Ohio

JOHN D. O'NEILL, University of Detroit, Detroit, Michigan

JAMES OSTERBURG, University of Cincinnati, Cincinnati, Ohio

DONALD PASSMAN, University of Wisconsin, Madison, Wisconsin

PAUL V. REICHELDERFER, Ohio University, Athens, Ohio

EDNA REITER, University of Cincinnati, Cincinnati, Ohio

FRANK L. SANDOMIERSKI, Kent State University, Kent, Ohio

M. S. SATYANARAYANA, Bowling Green State University, Bowling Green, Ohio

ROBERT L. SNIDER, Virginia Polytechnic Institute and State University, Blacksburg, Virginia

STUART A. STEINBERG, University of Toledo, Toledo, Ohio

RALPH STINEBRICKNER and his group, Berea College, Berea, Kentucky

D. TURNRIDGE, Kent State University, Kent, Ohio

JIM WHITEHEAD, Kent State University, Kent, Ohio

JOHN M. WORRELL, Jr., Ohio University, Athens, Ohio

CONTENTS

ALTERNATING IDENTITIES

S. A. Amitsur

Hebrew University, Jerusalem, Israel and University of California, San Diego

1. Introduction

The paper contains a survey of the use of alternating polynomials to the theory of rings with a polynomial identity and to the theory of central simple algebras. Most of the applications are new proofs to well known results, and a new central identity is computed.

Let $\Omega[x;y] = \Omega[x_1,x_2,\ldots;y_1,y_2,\ldots]$ be a free ring generated over a commutative domain Ω. At a certain stage we will assume that Ω is a universal domain so that it will contain all new indeterminates required.

Definition. A polynomial $p[x_1,\ldots,x_m;y_1,\ldots,y_\ell] = p[x]$ is said to be an alternating polynomial in $x_1,\ldots,x_m$ if it is homogeneous and multilinear in the x_i's, and vanishes if we substitute $x_i = x_j$ for some $i \neq j$.

This implies that

(1.1) $p[x_{\sigma(1)},\ldots,x_{\sigma(m)};y_1,\ldots,y_\ell] = (sg\sigma)p[x_1,\ldots,x_m;y_1,\ldots,y_\ell]$

for any permutation σ of $\{1,\ldots,m\}$. And that $p[x] = 0$ if some x_i is substituted by a linear combination of the rest $\{x_j\}$ $j \neq i$.

Among the alternating polynomials there is a classic one, namely the Capelli polynomial:

(1.2) $d_m[x] = d_m[x_1,\ldots,x_m;y_1,\ldots,y_{m-1}] =$

$$\Sigma(sg\sigma)x_{\sigma(1)}y_1x_{\sigma(2)}y_2\cdots y_{m-1}x_{\sigma(m)}$$

where the sum ranges over all permutations $\sigma \in S_m$.

The alternating identities are most useful in problems where elimination is a useful tool as they behave like determinants. In the following we develop some of the basic relations between alternating identities, compute a central alternating identity for prime rings. We survey some of the applications to ACC pi-rings. The Artin-Procesi theorem to which we provide another generalization, following a proof of Schelter. Some new applications to properties of central simple algebras are obtained by expressing them with the aid of generic matrices. The main properties are found in Amitsur [1], Razmyslov [2] and some of the formulas were obtained independently by Bergman. An important new application will be given in a forthcoming paper by Small-Amitsur.

2. Basic Formulas

With the aid of an alternating polynomial in m variables $x_1,\dots,x_m$ we can construct an alternating identity in m+1 variables. Namely,

Lemma 1. [1] If $p[x_1,\dots,x_m,y_1,\dots]$ is an alternating polynomial in $x_1,\dots,x_m$, then

(2.1) $$P[x_1,x_2,\dots,x_{m+1};y_0,y_1,\dots] = \sum_{i=1}^{m+1} (-1)^{i-1} p[x_1,x_2,\dots,\hat{x}_i,\dots,x_m,x_{m+1};y_1,\dots]y_0x_i$$

is an alternating polynomial in m+1 x_i's; where $\hat{x}_i$ denotes that x_i is omitted, and y_0 is a new indeterminates.

This is actually the development of an $(m+1) \times (m+1)$ determinants by its $m \times m$ minors.

The Capelli polynomials are the generators of all alternating identities in the following sense:

Lemma 2. [1] A polynomial $p[x;y]$ is alternating, if and only if

$$p[x_1,\dots,x_m,y_1,\dots] = \sum \alpha_\lambda q_{\lambda_0} d_m[x_1,\dots,x_m;q_{\lambda\Lambda},\dots,q_{\lambda_{m-1}}]q_{\lambda_m} \tag{2.2}$$

where q_{λ_j} are monomials in y's.

Another type of relations one obtains for alternating polynomials is by following the analogue of the multiplication theorem for determinants:

Let $\Omega x_1 +\dots+ \Omega x_m = X$ be the linear space generated by m x_i's over Ω and $T:X \to X$ be a linear transformation of X and $p[x_1,\dots,x_m;y_1,\dots]$ an alternating polynomial. Then

Lemma 3. $$p[Tx_1,\dots,Tx_m;y_1,\dots] = (\det T)p[x_1,\dots,x_m;y_1,\dots] \tag{2.3}$$

$$p[(\lambda - T)x_1,\dots,(\lambda - T)x_m;y_1,\dots] = \det(\lambda - T)p[x_1,\dots,x_m;y_1,\dots] \tag{2.4}$$

If the $c_j(T)$ are the coefficients of the characteristic polynomials of T, then:

$$c_j(T)p[x_1,\dots,x_m;y_1,\dots] = \sum_{i_1<i_2<\dots<i_j} p[x_1,\dots,Tx_{i_1},\dots,Tx_{i_j},\dots,x_m;y_1,\dots] \tag{2.5}$$

The proof for special case were given by Razmyslov [2], and also by G. Bergman. We follow the classical proof of the multiplication of determinants:

$$p[Tx;y] = \sum_{i,k} t_{i_1k_1}t_{i_2k_2} \dots t_{i_mk_m}p[x_{k_1},x_{k_2},\dots,x_{k_m};y_1,\dots].$$

When two k_i's are equal the corresponding term is zero since $p[x;y]$ is alternating and if $(k_1,\dots,k_m)$ is a permutation σ of $(1,\dots,m)$ then by rearranging the t's we get

$$\sum(\mathrm{sg}\,\sigma)t_{1\sigma(1)}t_{2\sigma(2)} \dots t_{m\sigma(m)}p[x_1,\dots,x_m;y_1,\dots] = (\det T)p[x;y].$$

Substituting $\lambda - T$ for T where λ is a new indeterminate, we obtain (2.4). Noticing that $|\lambda - T| = \sum \lambda^{n-j}(-1)^j c_j(T)$ we get (2.5) by comparing the coefficient of λ^{n-j} of both sides. We are in particular interested in the case $j = 1$; that is:

$$(2.6) \quad p[x_1,\dots,x_m;y_1,\dots]\operatorname{tr} T = \sum_{i=1}^{m} p[x_1,\dots,Tx_i,x_{i+1},\dots,x_m;y_1,\dots]$$

where tr T denote the trace of T.

3. Linear Dependence

Formula (2.1) in combination with (2.2) is very useful in providing conditions for linear dependence. Since an alternating polynomial $p[x_1,\dots,x_m;y,\dots]$ vanishes if two indeterminates are set to be equal, it will follow by a simple computation that:

Lemma 4. $p[x_1,\dots,x_m;y_1,\dots] = 0$ if the x_i are chosen from a submodule of a ring R generated by at most m-1 elements over the center of R.

This implies immediately that the matrix ring $M_n(K)$ over a commutative ring satisfies $p[x_1,\dots,x_{n^2+1},y_1,\dots] = 0$, and in particular $d_{n^2+1}[x;y] = 0$.

Also it will follow by lemma 1 and lemma 2:

Lemma 5. If R satisfies $d_{m+1}[x;y] = 0$ then for any alternating polynomial $p[x;y]$ in m-indeterminates, R satisfies:

$$(3.1) \quad p[x_1,\dots,x_m;y_1,\dots]x = \sum_{i=1}^{m} p[x_1,\dots,x_{i-1},x,x_{i+1},\dots,x_m;y_1,\dots]x_i.$$

This follows from the fact that by lemma 2, R satisfies all alternating polynomial identities in m+1 indeterminates and, in particular, the identity (2.1). Now (3.1) is obtained from this identity by setting $x_{m+1} = x$, $y_0 = 1$ and changing the sign of the term by moving x to the ith place.

For prime rings (and semi prime ring) of pid = n we can get a better result than lemma 5. Recall that a prime pi-ring R has pid = n (polynomial-identity-degree) if its minimal identity is of degree 2n, and equivalently, its ring of quotient Q(R) is a central simple algebra of dimension n^2 over its center $F = F(R)$, and hence has a splitting field K, i.e. $Q(R) \otimes K \simeq M_p(K)$.

For these rings we prove:

Theorem 6. If R is a prime ring of pid = n and center C, then R satisfies $d_{n^2+1}[x;y] = 0$, and a set of n^2 elements $r_1,\ldots,r_{n^2} \in R$ are C independent, if and only if, $d_{n^2}[r_1,\ldots,r_{n^2},s_1,\ldots,s_{n^2+1}] \neq 0$ for some $s_j \in R$. Furthermore, one can choose $\{s_j\}$ that $tr(d_{n^2}[r,s]) \neq 0$.

Proof. If the $\{r_j\}$ are C-dependent, then in Q(R) they lie in an F-space generated by less than n^2 elements, and by lemma 4 $d_{n^2}[r;s] = 0$ for all $s_j \in R$.

Conversely, if the $\{r_j\}$ are C-independent, they form a base of Q(R) and as well as base of $M_n(K)$ when identifies with $Q(R) \otimes K$. In particular $c_{ik} = \sum\alpha_{ikj}r_j$, $\alpha_{ikj} \in K$. If we look on this relation as a K-linear transformation, setting $Ar_{i+nk} = c_{ik+1}$, $i = 1,\ldots,n$, $k = 0,\ldots,n-1$ then $d_{n^2}[c_{11},\ldots,c_{nn};s] = |\alpha_{ikj}| d_{n^2}[r_1,\ldots,r_{n^2},s]$, and its sufficient, therefore, that the left side is non-zero and that for some $s_j \in M_n(K)$ it has a non-zero trace.

Indeed, choose for $\{x_1,\ldots,x_{n^2}\}$ the set $\{c_{12},\ldots,c_{1n},c_{21},\ldots,c_{2n},\ldots,c_{n1},c_{n2},\ldots,c_{nn},c_{11}\}$ and for $\{y_1,\ldots,y_{n-1}\}$ the set $\{c_{21},c_{31},\ldots,c_{n-1},c_{n2},\ldots, c_{n-1n},\ldots,c_{n-1n},c_{n1}\}$ in $d_{n^2}[x;y]$. The set of y's is so chosen that the monomial $x_1y_1x_2y_2 \ldots x_{n^2-1}y_{n^2-1}x_{n^2} = c_{11}$, but then this is the only term which is non-zero. Since in any monomial

$x_{\sigma(1)}y_1x_{\sigma(2)}y_2 \cdots y_{n^2+1}x_{\sigma(n^2)}$, the fact that $y_1x_{\sigma(2)}y_2 \neq 0$ implies that $x_{\sigma(2)} = c_{13} = x_2$, and so on, $x_i = x_{\sigma(i)}$ for $i = 2,\ldots,n^2-1$. The only remaining terms which can be chosen are $\{c_{12},c_{11}\}$ but clearly only $x_{\sigma(1)} = x_1 = c_{12}$, $x_{\sigma(n^2)} = x_{n^2} = c_{11}$ is the only possibility to yield a non-zero term. Thus, under this substitution $d_{n^2}[x;y] = c_{11}$ and $\operatorname{tr} d_{n^2}[x;y] = 1$. Q.E.D.

Let $X_1,\ldots,X_{n^2}$ be n^2 generic matrices, i.e. $X_\lambda = \sum\xi_{ik}^\lambda c_{ik}$ with ξ_{ik}^λ are independent commutative indeterminates. A condition for the independence of the X_i's can be expressed by a $n^2 \times n^2$ determinant obtained by writing the entries of X_λ is one row $(\xi_{11}^\lambda \ldots \xi_{1n}^\lambda\xi_{21}^\lambda \ldots \xi_{2n}^\lambda \ldots \xi_{n1}^\lambda \ldots \xi_{nn}^\lambda)$. Denote the corresponding determinant by $\det(X_1,\ldots,X_{n^2}) = \det X$. Then, a set of n^2 matrices $\{A_\lambda\}$ are linearly independent if and only if $\det(A_1,\ldots,A_{n^2}) \neq 0$. Thus $\det X \neq 0$ is a necessary and sufficient condition for independence. The connection between this criterion and the criterion of theorem 6 is given by the following

$$(3.2) \qquad p[X_1,\ldots,X_{n^2};Y_1,\ldots] = \det(X_1,\ldots,X_{n^2})q[Y_1,\ldots]$$

where $q[Y_1,\ldots]$ is a matrix whose entries depend only on the entries of the Y's and do not involve any of the ξ_{ik}^λ.

The proof follows from (2.3). Consider the linear space generated by the matrix units c_{ik} over the domain Ω, which we assume to contain all entries of the X's and the Y's. If we set $c_{i+kn} = c_{ik+1}$ $i = 1,\ldots,n; k = 0,1,\ldots,(n-1)$ and $Tc_\lambda = X_\lambda = \sum\xi_{ik}^\lambda c_{i+kn}$ then it follows from (2.3) that $p[x;Y] = p[Tc,Y] = (\det T)p[c,Y]$. Clearly, $\det T = \det(X_1,\ldots,X_{n^2})$ and $p[c,Y] = q[Y]$ has the required properties.

4. A Central Identity

We apply the preceding formulas to obtain an explicit formula for a multilinear alternating identity in $2n^2$ indeterminates which is a central identity, and does not vanish in any matrix ring $m_n(K)$, independent of the characteristic.

Theorem 7. In any matrix ring $M_n(\Omega)$: set $y_0 = 1$ then

$$\operatorname{tr}(d_{n^2}[x_1,\ldots,x_{n^2};y_1,\ldots,y_{n^2-1}])\cdot \operatorname{tr} y_{n^2} =$$

$$(4.1)\quad = \sum_{j=1}^{n^2} (-1)^{j(n^2-1)} y_j d_{n^2}[x_1,\ldots,x_{n^2};y_{j+1},\ldots,y_{n^2-1}, y_0,y_1,\ldots,y_{j-2},(y_{j-1}y_{n^2})]$$

is an alternating central identity (i.e. its values belong to the center and it is not identically zero).

Proof. Let the x's, y's be generic matrices, as well as a,b and consider the transformation T given by $Tx_\lambda = ax_\lambda b$ which is linear since the $\{x_\lambda\}$ will be a base of $M_n(\Omega)$ if Ω contains the field of all rational functions in the entries. The characteristic polynomial $|\lambda - T| = f_T(\lambda)$ is independent of the base, so if we choose them to be $\{c_{ik}\}$ then one observes that the matrix corresponding to T will be $a \times b^t$ where "×" is the Kronecker product of the matrices, and b^t is the transpose of b. In particular, we have: $\operatorname{tr} T = \operatorname{tr}(a \times b^t) = (\operatorname{tr} a)(\operatorname{tr} b)$. Hence we obtain by (2.6): (Razmyslov [2], Bergman)

$$(4.2)\quad d_{n^2}[x;y]\operatorname{tr} a \operatorname{tr} b = \sum_{i=1}^{n^2} d_{n^2}[x_1,\ldots,ax_ib,\ldots,x_{n^2};y_1,\ldots,y_{n^2-1}].$$

Write the sum in the form $\sum \mu_j a v_j b w_j$.

Taking the trace of both sides and noting that $\operatorname{tr}(uv) = \operatorname{tr}(vu)$:

$$\operatorname{tr}[d_{n^2}(x,y)\operatorname{tr} a \operatorname{tr} b] = \operatorname{tr} d_{n^2}[x,y]\operatorname{tr} a \operatorname{tr} b = \operatorname{tr}[\operatorname{tr} d_{n^2}[x,y](\operatorname{tr} a)\cdot b]$$

$= tr(\Sigma u_j a v_j b w_j] = \Sigma tr(u_j a v_j b w_j) = \Sigma tr(w_j u_j a v_j b) = tr[(\Sigma w_j u_j a v_j)b].$

The tr(uv) is regular bilinear form, i.e. tr(ua) = tr(va) for every a implies that u = v. In our case it yields that

$$(tr\ d_{n^2}[x,y])(tr\ a) = \Sigma v_j b w_j u_j = f[x,y,b].$$

Thus, it is a polynomial in the x's, y's and b, but central as a product of traces and not identically zero since $tr\ d_{n^2}[x,y] \neq 0$, $tra \neq 0$ by theorem 6. It is alternating since $d_{n^2}[x,y] = 0$ if two $x_i = x_j$, $i \neq j$. We, in fact, can compute $f[x,y,y_{n^2}]$ (putting $a = y_{n^2}$, also $y_0 = 1$)

$$\sum_{i=1}^{n^2} d_{n^2}[x_1,\ldots,y_{n^2}x_i b,\ldots,x_{n^2};y_1,\ldots] =$$

$$\sum_{i=1}^{n^2}\sum_{j=1}^{n^2}\sum_{\sigma(j)=i}(sg\sigma)x_{\sigma(1)}y_1 \cdots y_{j-1}y_{n^2}x_i b y_j \cdots y_{n^2-1}x_{\sigma(n^2)}$$

Thus, the preceding computation yields

$$f[x,y,y_{n^2}] = \sum_{i=1}^{n^2}\sum_{j=1}^{n^2}\sum_{\sigma(j)=i}(sg\sigma)y_j x_{\sigma(j+1)} \cdots y_{n^2-1}x_{\sigma(n^2)}x_{(1)}y_1 \cdots$$

$$x_{\sigma(j-2)}y_{j-1}y_{n^2}x_i =$$

$$\sum_j\sum_{\sigma^k}(sg\sigma)y_j x_{\sigma*(1)}y_{j+1}x_{\sigma*(2)} \cdots y_{n^2-1}x_{\sigma*(n^2-j)}x_{\sigma*(n^2-j+1)}y_1 \cdots$$

$$y_{j-1}y_{n^2}x_{\sigma*(n^2)}$$

where $\sigma^* = \begin{pmatrix} 1 & 2 & \ldots & n^2-j & n^2-j+1 & \ldots & n^2 \\ \sigma(j+1) & \sigma(j+2) & & \sigma(n^2) & \sigma(1) & & \sigma(j) \end{pmatrix}$ which is obtained from σ by multiplication by the cycle $(12 \ldots n^2)^j$. Thus, $sg\sigma^* = (-1)^{(n^2-1)j}sg\sigma$. So finally, we obtain by setting $y_0 = 1$:

$$tr\ d_{n^2}[x,y]tr\ y_{n^2} = \sum_{j=1}^{n^2}(-1)^{j(n^2-1)}y_j d_{n^2}[x_1,\ldots,x_{n^2};$$

$$y_{j+1},\ldots,y_{n^2-1},y_0,y_1,\ldots,y_{j-1}y_{n^2}] \qquad \text{i.e. } (4.1)$$

Another fact follows from (4.2):

Corollary 8. The C-module generated by all values $d_{n^2}[x;b]$ $b_j \in R$ of a prime PI-ring of pid = n, is the two sided ideal generated by all these values.

Indeed, set b = 1 in (4.2) and we get:

$$n\,tr(a)d_{n^2}[x,y] = \sum_{i=1}^{n^2} d_{n^2}[x_1,\dots,ax_i,\dots,x_{n^2};y_1,\dots] =$$

$$= a\sum_{i=1}^{n^2}\sum_{\sigma(1)=i} sg\sigma x_i y_1 x_{\sigma(2)} \cdots y_{n^2-1}x_{\sigma(n^2)} +$$

$$\sum_{i=1}^{n^2}\sum_{j=2}^{n^2}\sum_{\sigma(j)=1} sg\sigma x_{\sigma(1)} \cdots (y_{i-1}a)x_i \cdots y_{n^2-1}x_{\sigma(n^2)}$$

$$= ad_{n^2}[x_1,\dots,x_{n^2};y_1,\dots,y_{n^2-1}] +$$

$$\sum_{j=1}^{n^2-1} d_{n^2}[x_1,\dots,x_{n^2};y_1,\dots,y_{j-1}a,\dots,y_{n^2-1}].$$

Hence, $ad_{n^2}[x,y] = \sum_{j=1}^{n^2} d_{n^2}[x;y_1,\dots,y_ja,\dots,y_{n^2-1}] - n\,tr\,(a)d_{n^2}[x,y].$

The fact that $\sum \alpha d_{n^2}[x;b]$ is also a right ideal is obtained similarly by setting b = 1 in (4.2).

5. Applications to Structure of PI-Rings

Let R be a K-algebra, K a commutative ring. Prime rings satisfy $d_{n^2+1}[x;y]$ and have a central identity $\delta[x;y]$ which is alternating in n^2 indeterminates and which does not vanish identically in any $M_n(\overline{K})$, $\overline{K}$ homomorphic image of K. In (1) we consider such rings, and denoted by Δ the ideal generated by all values $\delta[a;b]$, a_i, $b_j \in R$. Among others, the following results have been shown in [1]:

Theorem 9. If R has the preceding properties, and satisfies ACC on two sided ideals, then $R/(0:\Delta)$ is both right and left noetherian.

The case R prime is due to Cauchon.

Theorem 10. (Artin-Procesi) R as before, $1 \in R$, pid $R = n$ and for every maximal ideal M of R the quotient ring R/M is central simple of dimension n^2 then R is an Azumaya algebra of rank n^2.

We present here a slight generalization of this theorem, by following a proof of Schelter to theorem 10, which we state in the following two lemmas:

Lemma 11. Let R be as before, and $u \in \Delta$, then $uR \subseteq$ in a finitely generated C-module.

Proof. Let $u = \sum_j \delta[a_{j1},\ldots,a_{jn^2};b_{j1},\ldots]r_j$, then for every $r \in R$ we obtained by (3.1):

$$ur = \sum\delta[a_{j\nu},b_{j\nu}]r_jr = \sum_{j,\nu}\delta_{j\nu}[r_jr]a_{j\nu} \in \sum Ca_{j\nu}$$

where $\delta_{j\nu}[r_jr] = \delta[a_{j1},\ldots,a_{j\nu-1},r_jr,a_{j\nu+1},\ldots,a_{jn^2};b_{j1},\ldots]$. Thus $uR \subseteq \sum Ca_{j\nu}$ as required.

Lemma 12. Let $\Phi: R \otimes R^\circ \to \mathrm{Hom}_c(R,R) = \mathrm{End}_c(R)$ given by $\Phi(x\otimes y^\circ)[z] = xzy$. Then $\mathrm{IM}(\Phi) \supseteq \Delta\mathrm{End}_c(R)$, $\mathrm{Ker}\Phi \subseteq (0:\Delta)$ in $\mathrm{End}(R)$.

Proof. For every $\delta[a_1,\ldots,a_{n^2},b_1,\ldots] = \delta[a,b]$ and $\tau \in \mathrm{Hom}_c(R,R)$:

$$(\delta[a,b]\tau)(x) = \tau(\delta[a,b]x) = \tau(\sum\delta_j[x]a_j) = \sum\delta_j[x]\tau a_j$$

where $\delta_j[x] = \delta[a_1,\ldots,a_{j-1},x,a_j,\ldots,a_{n^2};b_1,\ldots]$, and $\delta x = \sum\delta_j[x]a_j$ by (3.1). Express now each $\delta_j[x] = \sum u_{j\lambda}xv_{j\lambda}$, and so

$$[\delta[a,b]\tau](x) = \sum u_{j\lambda}xv_{j\lambda}\tau a_j = \Phi[\sum u_{i\lambda}\otimes(v_{j\lambda}\tau a_j)^\circ](x) \quad \text{i.e. } \delta\tau \in \mathrm{Im}\Phi.$$

Now $\Phi(\sum u_i\otimes v_i^\circ) = 0$ means that for every $x \in R$ $\sum u_ixv_i = 0$. So if $\delta = \delta[a,b] \in \Delta$ then by (3.1) as before:

$$\delta\sum u_i\otimes v_i^\circ = \sum\delta[a,b]u_i\otimes v_i^\circ = \sum_{ij}\delta_j[u_i]a_j\otimes v_i^\circ = \sum_j a_j\otimes \sum_i\delta_j[u_i]v_i^\circ$$

$$= \sum_j a_j\otimes(\sum_i\sum_\lambda u_{j\lambda}u_iv_{j\lambda}v_i)^\circ = \sum a_j\otimes\sum_\lambda(\sum u_iv_{i\lambda}v_i)^\circ u_{i\lambda}^\circ = 0.$$

That is $\mathrm{Ker}\Phi \subseteq (0:\Delta)$ in End(R).

An immediate corollary of both lemmas is the proof of theorem 9:

Proof of Theorem 10. Under the conditions of the theorem $\Delta = R$, since for any maximal ideal M containing Δ, R/M will satisfy $\delta[x_1,x_2,\ldots,x_{n^2},y] = 0$ and hence will be of pid < n, which contradicts our assumption.

Now choose u = 1 in lemma 11, and obtain that R is a finitely generated C-module. Also since $1 \in \Delta$, it follows by lemma 12 that Φ is an isomorphism and these are the conditions that R is an Azumaya algebra. Q.E.D.

The computations also prove:

Corollary 13. If R as in the beginning of the section and $\sum u_i x v_i = 0$ for every $x \in R$, then for every $\delta[a_1,\ldots,a_{n^2-1},z;b_1,\ldots]$:

$\sum\delta[a_1,\ldots,a_{n^2-1},u_i,b_1,\ldots]v_i = 0$. i.e. a linear dependence $\sum\delta[u_i]v_i = 0$ of the v_i over C.

If R is prime, then for some a_j, $\delta[u_i] \neq 0$ for some $u_i \neq 0$.

The computation achieving this relation appears in the computation of $\mathrm{Ker}\Phi$.

A final application for structure theorem is a different proof of the following result of Razmyslov.

Theorem 14. Let Δ_c be the ideal generated by all $\delta[a;b]$ in the center C of a ring of pid = n with the properties of the beginning of this section. Then R has an ideal A integral over Δ_c, and $A \supseteq M_{n-1}(R)$ where the latter denotes the ideal generated by all values of f[x], and f[x] ranges over all identities of $n-1 \times n-1$ matrices.

Proof. By classical arguments about radicals, it follows that R contains an ideal A maximal with respect to the property that all its elements are integral over Δ_c. Furthermore, A is a semi-prime ideal.

Indeed, if L/A is the lower radical of R/A, then L/A is nil and hence also L contains only integral elements over Δ_c, but A is maximal. Hence $L = A$.

We now prove that $A \supseteq \Delta_c R = \Delta$. Indeed for every $\delta[a,b]$ we have by (3.2): $\delta[a,b]ra_i = \sum \delta_k[ra_i]a_k$, δ, $\delta_k \in \Delta_c$. Hence if $f[\lambda] = |\lambda - (\delta_k[ra_i])|$ the characteristic polynomial of the matrix of the coefficients then $f[\delta x]a_i = 0$. Hence $f[\delta x]\delta x = \sum f[\delta x]\delta_i[x]a_i = 0$ and so δR is an ideal containing only integral elements. Thus, $A \supseteq \delta R$ and hence $A \supseteq \Delta_c R$.

Finally, for every prime ideal $P \supseteq A$, $P \supseteq \Delta_c$, and so R/P satisfies $\delta[x;y] = 0$, which implies that pid $R/P < n$. Consequently, $P \supseteq M_{n-1}(R)$. Now A is semi prime, so $A = \cap P \supseteq M_{n-1}(R)$. Q.E.D.

6. Ring of Generic Matrices

Let $\Omega[X_1,\ldots,X_r]$ be the ring of $n \times n$ generic matrices; $\Omega[\xi]$ be the commutative ring of polynomial in the entries ξ_{ik}^{λ} of the matrices $\{X_\lambda\}$.

We then have $\Omega[X_1,\ldots,X_n] \subset \Omega[\xi][X_1,\ldots,X_n] \subset M_n(\Omega[\xi])$. If Ω is a domain then $\Omega[X]$ is a domain but $\Omega[\xi][X]$ is never a domain. In fact:

Theorem 15. For every ideal $I \triangleleft \Omega[X]$ the ideal $\Omega[\xi]I$ in $\Omega[\xi][X]$ contains a matrix ring over an ideal in $\Omega[\xi]$. In particular $\Omega[\xi][X]$ will contain $M_n(\Delta)$ where Δ is the ideal in the center C of $\Omega[X]$ generated by all central alternating identities of $M_n(\Omega)$.

Proof. Let c_{ik} be a matrix base of $M_n(\Omega[\xi])$ and $\delta[x_1,\ldots,x_{n^2};y]$ an alternating central polynomial, then by (3.2) $\delta[a,b]c_{ik} = \delta[a_1,\ldots,a_{n^2};b,\ldots]c_{ik} = \sum \delta_k[c_{ik}]a_i$.

If $\{a_i\}$ are taken from an ideal I, then the left side ranges over all matrices $M_n(\Delta[I])$, where $\Delta[I]$ is the set of all values of all

$\delta[a,b]$ for all $a_i \in I$ (and it is $\neq 0$). The right side $\subseteq \Omega[\xi]I$, since $a_i \in I$ and $\delta_k[c_{ik}]$ are central elements in $M_n(\Omega[\xi])$ and so $\in \Omega[\xi]$. If $I = \Omega[X]$ we obtain also the second part of the theorem.

The preceding clearly imply a theorem of Procesi.

Corollary 16. The field of rational functions $\Omega(\xi)$ split the ring of quotients of $\Omega[X]$.

7. Applications to Central Simple Algebras

In this last section an example is given to show how alternating identities can be obtained to prove generic properties of central simple algebras of dim n^2.

In the setting of (2.4), consider the transformation T gives: $Tx_i = ax_i - x_i a$ for a fixed matrix a. The matrix of this transformation with respect to the matrix base $\{c_{ik}\}$ is readily seen to be $a \times 1 - 1 \times a^t$ where $\times$ is the Kronecker product of matrices, and a^t the transpose matrix. So (2.4) yields

$$(6.1) \qquad p[\lambda x_i - (ax_i - x_i a);y] = \det(\lambda - (a \times 1 - 1 \times a^t)p[x;y].$$

If a is a diagonal matrix $a = \mathrm{diag}\{\xi_1,\ldots,\xi_n\}$ then $\det|\lambda - (a \times 1 - 1 \times a^t)| = \prod_{c_{ik}} [\lambda - (\xi_i - \xi_k)]$. Hence the coefficient of λ^n is $\prod_{i \neq k} (\xi_i - \xi_k)$ which is the square of discriminant D(a) of the characteristic polynomial of a. This implies that for generic matrix z, y's (which can be diagonalized) and x's, y's we have the relation:

$$(6.2) \qquad p[x;y]D(z)^2 = g_p[x,y,z]$$

and g[x,y,z] is a polynomial in the x's, y's and in z.

Now $g_p[x;y\cdot z] \neq 0$ in $M_n(F)$ for any field F, since alternating identity $p[x,y] \neq 0$ in $M_n(F)$. e.g. $d_{n^2}[x,y]$, and in a matrix.

$M_n(F)$ for any field F where $z = a$ for which $D(a) \neq 0$. For infinite field the proof is trivial, and for finite field choose a matrix a whose characteristic matrix is an irreducible polynomial of degree n.

But then $g_p[x,y,z] \neq 0$ in any prime ring R of pid = n. In particular if R is a central simple algebra of dimension n^2 over a field K the element $r \in R$ for which there exists $g_p[a,b,r] \neq 0$, has a discriminant $D(r) \neq 0$; i.e. r is separable. Consequently, we have shown a classical result, with the aid of this identity:

Corollary 17. A central simple algebra R of dimension n^2 over a field K contains a separable element $r \in R$, and thus $[K(r):K] = n$, and, therefore, $K(r)$ is a maximal commutative separable subring of R.

Other examples of this type which have applications of a different form will appear in a forthcoming paper of L. Small and the author.

References

[1] S. A. Amitsur, "Identities and linear dependence," Israel J. Math., 22 (1975), 127-137.

[2] Ju. P. Razmyslov, "The Jacobson radical in PI-algebras," Algebra i Logika, 13 (1974), 327-360.

THE THEORY OF GENERALIZED POLYNOMIAL IDENTITIES

Louis Halle Rowen

Bar Ilan University
Ramat Gan, Israel

§0. Introduction

One of the most fruitful subjects in recent years has been the theory of polynomial identities of rings (called PI-theory), which has provided a most natural framework for expanding on and generalizing the theory of finite dimensional algebras (which can easily be seen to satisfy polynomial identities). One method of developing the PI-theory has been to study the n × n matrix algebra $M_n(F)$ in terms of polynomial identities, and then to study a semiprime ring by embedding it in a direct product of matrix algebras satisfying the same identities. This procedure "works" because of Kaplansky's theorem [17], which says that every primitive PI-ring is central simple. A natural question is, if we weaken the PI-theory to admit coefficients of the polynomials from the ring itself, can we still obtain a structure theorem for primitive rings satisfying these "generalized" polynomial identities? Amitsur [1] found such a characterization, which is related to $\mathrm{End}_F M$, the endomorphism ring of a vector space M over a field F. It turns out that one can build a whole theory of rings with generalized polynomial identities, based first on studying $\mathrm{End}_F M$ and then embedding semiprime rings into direct products of endomorphism rings. In other words, the focal point is generalized, from $M_n(F)$ in the PI-theory, to $End_F M$ (which of course is a matrix algebra if M is finite dimensional).

The object of this expository paper is to build such a theory of generalized polynomial identities, in a manner which has not previously appeared in the literature but which unites results of Amitsur, Martindale, Jain, and Rowen. Immediately there is a question of how one chooses the coefficients for the generalized polynomials. This was originally an obstruction in the theory, but we use here the concept of "algie," generalizing "algebra," which enables us to treat generalized (polynomial) identities in a varietal context (cf. §2). $End_F M$ is studied in §3, and the structural embedding theorems, which rely largely on results of Martindale and Amitsur, are given in §4. The results of §3 and §4 yield the general structure theorems in §5.

An unexpected bonus of the theory of generalized identities has been its successful applications to problems in PI-theory, some of which are given in §6. Martindale's "central closure" of a prime ring is studied in §7; sometimes this construction is useful in gaining additional useful knowledge about prime rings. The emphasis of this exposition is on rings <u>with</u> 1, but §8 shows how one can carry most of the theory through for rings without 1.

The theory of generalized identities has been very successful in handling questions on rings with involution, and in some ways the more flexible choice of coefficients makes it easier to prove results more extensive than the theorems for PI-rings with involution. We do not handle involutions here, but remark that most of the theorems in this paper have generalizations in the theory of generalized identities of rings with involution.

§1. A Brief Survey of Parts of Classical PI-theory

Before we embark on the theory of generalized identities, let us glance briefly at the theory of polynomial identities, to see just which ideas we should be generalizing. Throughout this paper, rings will be associative with 1, unless otherwise indicated. Suppose Ω is a commutative ring, and let $\Omega\{X\} = \Omega\{X_1,X_2,\ldots\}$ be the free Ω-algebra generated by a countable set of noncommuting indeterminates $X_1,X_2,\ldots$ over Ω. The elements of $\Omega\{X\}$ are called polynomials; there is a canonical way of writing a polynomial as a sum of its monomials. Given a polynomial $f(X_1,\ldots,X_m)$, define $R(f) = \{f(r_1,\ldots,r_m) \mid \text{all } r_i \text{ in } R\}$ = the set of evaluations of f in R. (This can be viewed formally as the set of images of f taken under all Ω-homomorphisms from $\Omega\{X\}$ to R.) We say f is an identity of R if $R(f) = 0$. If the coefficients of f are all in $\mathbb{Z}$, call f a classical polynomial; if one of the coefficients of f is ± 1, f is a strong classical polynomial. If there exists a coefficient of f which does not annihilate R, we say f is R-proper. R is a PI-ring if some strong classical polynomial f is an identity of R; we also say that f is a strong identity of R.

It is immediate that $X_1X_2-X_2X_1$ is a strong classical polynomial, and is an identity of R if and only if R is commutative; thus PI-rings are generalizations of commutative rings. On the other hand, define the standard polynomial $S_k(X_1,\ldots,X_k) = \Sigma(\operatorname{sg}\pi)X_{\pi 1}\cdots X_{\pi k}$, summed over all permutations π of $(1,\ldots,k)$. S_k is multilinear (i.e. linear in each indeterminate), and alternating (i.e. $S_k(\ldots,X_i,\ldots,X_i,\ldots) = 0$). It follows that S_k is an identity of every algebra of dimension $\leq k-1$ over a field, so these are all PI-rings. (In fact, a theorem of Amitsur-Levitzki [6] shows that S_{2d} is an identity of every central simple algebra of dimension $\leq d^2$ and is the identity with smallest degree.)

A nice property of S_k is the formula

$$S_k = \sum_{i=1}^{k} (-1)^{i+1} X_i S_{k-1}(X_1,\ldots,X_{i-1},X_{i+1},\ldots,X_k)$$
$$= \sum_{i=0}^{k-1} (-1)^{i} S_{k-1}(X_1,\ldots,X_{k-i-1},X_{k-i+1},\ldots,X_k)X_{k-i}.$$ Thus, if

$X_kS_{k-1}(X_1,\ldots,X_{k-1})$ or $S_{k-1}(X_1,\ldots,X_{k-1})X_k$ is an identity of a ring R, then S_k is an identity of R.

A ring is primitive if it has a faithful, irreducible (left) module.

Theorem A (Kaplansky [17]). If R is primitive and f is a strong identity of R, of degree d, then R is central simple, of dimension $\leq [d/2]^2$ over its center.

These facts suggest the study of finite-dimensional algebras by proving theorems about all PI-rings, with the suggestion that it may be possible to employ techniques arising in the theory of commutative rings. Over the last 25 years, this idea has proved fantastically successful. One major accomplishment was the streamlining of Jacobson's structure theory of rings, as follows:

A subalgebra R_1 of an algebra R is nil if every element of R_1 is nilpotent. Let $\mathrm{Nil}(R) = \cup\{\text{nil ideals of } R\}$, a nil ideal of R. Call R prime if the product of any two nonzero ideals is nonzero; R is semiprime if R has no nonzero nilpotent ideals. If R is a semiprime algebra then R is a subdirect product of prime algebras, each of which satisfies all identities of R. If $V \subseteq R$, define
$\mathrm{Ann}V = \{r \in R \mid Vr = 0\}$.

Theorem B (Levitzki-Amitsur, cf. [13, p. 232]). If R is a semiprime PI-ring, having a strong identity of degree d, then, for every nil subring R_1 of R, we have $R_1^d = 0$.

Corollary C. If R is a PI-ring, then the upper and lower nilradicals coincide, and Nil(R) contains every nil left ideal (and every nil right ideal) of R.

Amitsur proved, for any associative ring R, if Nil(R) = 0, then the polynomial ring $R[\lambda]$ has Jacobson radical 0, and is thus a subdirect product of a set of primitive homomorphic images $\{R_\lambda | \lambda \in \Gamma\}$ of $R[\lambda]$. Moreover, an easy argument shows that every multilinear identity of R is an identity of $R[\lambda]$. Thus, using theorem A and corollary C, we have the very important structural result that if R is a semiprime PI-ring, then $R[\lambda]$ is a subdirect product of simple artinian algebras R_γ each satisfying all the multilinear identities of R. Using a certain filter on the R_γ (which we shall describe in §3) and forming the ultraproduct of the R_γ (under this filter), Amitsur [3] obtained the following extension of a theorem of Posner [23]:

Theorem D. If R is a prime PI-ring then R has a simple artinian classical left and right ring of quotients R', which satisfy all the identities of R, and R' = Z(R')R.

One question which arises in this theory, is what assumption must we make on an identity of R, to conclude that R is a PI-ring? This was answered in a number of papers (cf. [2],[3],[4]) by Amitsur, and we summarize some of his results. Call a polynomial $f(X_1,\ldots,X_m)$ R-proper if $\omega R \neq 0$ for some coefficient ω of f; call f strongly R-proper if, for each $r \neq 0$ in R, we can find a coefficient ω_r of f such that $\omega_r r \neq 0$; call f R-strong if f is $\bar{R}$-proper for every nonzero homomorphic image $\bar{R}$ of R. Note, in case R is prime, that R-proper is the same as strongly R-proper; indeed, if $\omega r \neq 0$, then, for all $r' \neq 0$ in R, $\omega r'Rr = r'R\omega r \neq 0$, implying $\omega r' \neq 0$.

Theorem E (Amitsur). The following conditions are equivalent for a semiprime algebra R:

(i) R satisfies a strongly R-proper identity of degree d; (ii) R satisfies an R-strong identity of degree d; (iii) $S_{2[d/2]}$ is an identity of R.

To generalize theorem E to arbitrary algebras, Amitsur [4] used a trick which we later shall need in a slightly more general form.

Theorem 1.1. Suppose $\mathcal{R}$ is a class of rings, closed under taking of direct powers, such that, for any R in $\mathcal{R}$, R/Nil(R) is a PI-ring. Then every R in $\mathcal{R}$ is a PI-ring. Moreover, if f is an identity of R/Nil(R) for each R in $\mathcal{R}$, then, for any R in $\mathcal{R}$, some power of f (depending on R) is an identity of R.

Proof. Let $R^{(k)}$ = {k-tuples of elements in R}, indexed by Γ_k, and let $\Gamma = \cup\{\Gamma_k | k \geq 1\}$. For each γ in Γ, let R_γ be a copy of R, and form $R' = \pi\{R_\gamma | \gamma \in \Gamma\}$. By hypothesis, R'/Nil(R') is a PI-ring; hence some strong classical polynomial $g(X_1,\ldots,X_m)$ is an identity of R'/Nil(R'). Choose h_i in R', $1 \leq i \leq m$, such that, for γ in Γ_m corresponding to $(r_{1\gamma},\ldots,r_{m\gamma})$, $h_i(\gamma) = r_{i\gamma}$. Since $g(h_1,\ldots,h_m) \in \text{Nil}(R')$, we have $g(h_1,\ldots,h_m)^k = 0$ for some k. Hence $0 = g(h_1(\gamma),\ldots,h_m(\gamma))^k = g(r_{1\gamma},\ldots,r_{m\gamma})^k$. But $(r_{1\gamma},\ldots,r_{m\gamma})$ range over $R^{(m)}$, so $g(r_1,\ldots,r_m)^k = 0$ for all $r_1,\ldots,r_m$ in R. Therefore g^k is an identity of R.

Corollary 1.2. If R satisfies a strongly R-proper identity of degree d, or if R satisfies an R-strong identity of degree d, then $(S_{2[d/2]})^k$ is an identity of R. (In particular, R is a PI-ring.)

The smallest d such that, for suitable k, $(S_{2d})^k$ is an identity of R, is called the PI-degree of R. Now since $1 \in R$, there is another characterization of "proper" which shall prove very useful.

Remark 1.3. Write a polynomial f as the sum of its monomials h_i, $1 \leq i \leq u$. Clearly f is R-proper iff $R(h_i) \neq 0$ for some h_i; f is strongly R-proper iff $\mathrm{Ann}(\sum_{i=1}^{u} R(h_i)) = 0$.

Up to this point, however, there lacked a method of applying techniques of commutative rings to PI-algebras. Let $Z(R)$ denote the center of R, and call a classical polynomial $f(X_1,\ldots,X_m)$ R-central if $0 \neq R(f) \subseteq Z(R)$. A major accomplishment was the discovery by Formanek [11], and later by Razmyslov [24], of a strong classical polynomial, with constant term 0, which is $M_n(\Omega)$-central for every commutative ring Ω; the polynomial depends only on n. Using the structure described above, Rowen (cf. [25]) proved that every nonzero ideal of a semiprime PI-ring intersects the center nontrivially, and it follows easily that if we localize a prime PI-ring R at $(Z(R) - \{0\})$, we obtain a central simple algebra R', called the ring of central quotients of R, which satisfies all multilinear identities of R. This must be the R' of theorem D. (This was discovered independently by many people.) In fact, R' satisfies every identity of R, as seen via [25, theorem 3.1].

It is often useful to speak of identities of rings without 1. Such a concept is intuitively clear, and we shall not formalize it here, since a thorough development in the case of generalized identities will be given in §8.

§2. Generalized Polynomial Identities - Basic Properties

In §1, the study of PI-rings was reduced, through some structure theory, to the study of primitive PI-rings, which by Kaplansky's theorem are central simple; much effort in PI-theory has been spent in examining central simple algebras, and, in particular, matrix algebras, through the use of polynomials. Any worthwhile generalization of PI-theory should provide a way of examining at least one of the two natural generalizations of matrix algebras: simple artinian rings (which are matrix rings over division rings) or rings of endomorphisms of arbitrary vector spaces over a field. The theory of generalized polynomial identities treats the latter situation. Well, consider a vector space V over a field F, and let $R = End_F(V)$. If $dim_F V$ is not finite, then there are no R-proper identities. However, if we choose y in V and define e in R, by e(y) = y and e(y') = 0 for all y' F-independent of y, then $F \approx eRe$, implying $(er_1e)(er_2e)-(er_2e)(er_1e) = 0$ for all r_1, r_2 in R. Since $e^2 = e$, we might say $eX_1eX_2e-eX_2eX_1e$ vanishes for all substitutions in R. Thus, we have something like a polynomial identity, except that its "coefficients" do not necessarily commute with R. Hence, one is led to replace the coefficient ring Ω (of §1) by a noncommutative ring W, over which R is not quite an algebra.

Accordingly, let W be a ring, and suppose we are given a ring R which is a W-W bimodule. There is a canonical map $\psi: W \to R$, given by $\psi(w) = w \cdot 1$. We call R a W-<u>algie</u> if ψ is a ring homomorphism which preserves both module multiplications, i.e. $\psi(w)r = wr$ and $r\psi(w) = rw$ for all w in W, all r in R. If $\ker(\psi) = 0$, say R is W-<u>faithful</u>. We shall henceforth assume that R is a W-algie. Note that ideals of R are closed under left and right multiplication by elements of W, and are thus "W-algie ideals." Hence ring-theoretic terminology is consistent with "algie-theoretic" terminology. In [27] it was also required in algies that $\psi(Z(W)) \subseteq Z(R)$. This condition has two major advantages: (i) if W is commutative, then W-algies are merely W-algebras, and (ii) our definitions are seen to be equivalent to previous definitions used in [1], [15], and [19]. However, the theory runs smoother without this condition, because certain universal considerations are simplified. Note that if $R \subseteq R'$ as rings then R' is also a W-algie, with operations given by $wr' = (w1)r'$ and $r'w = r'(w1)$. Also, $W\cdot 1 \subseteq R$ is a W-algie.

There is a free W-algie W{X}, the free product (over $\mathbb{Z}$) of W and $\mathbb{Z}${X}. The elements of W{X} are called generalized polynomials, and can be written (not necessarily in a canonical manner) as sums of monomials of the form $h = w_1X_{i_1}w_2X_{i_2} \cdots w_mX_{i_m}w_{m+1}$. The w_i are called coefficients and the X_i are called indeterminates; $X_{i_1}X_{i_2} \cdots X_{i_m}$ is called the fingerprint of h. Given a generalized polynomial f, choose a way of writing f as a sum of the smallest possible number of monomials; these will be called the monomials of f, and we will write $f(X_1,\ldots,X_m)$ to denote that $X_1,\ldots,X_m$ are the indeterminates occurring in the monomials of f. A generalized monomial of f is the sum of the monomials of f with a given fingerprint. It is easy to see that there is a canonical, unique way of writing any generalized polynomial as a sum of its generalized monomials.

As an example of these ideas, consider $f(X_1) = w_1X_1w_3 + w_2X_1w_3 + w_2X_1w_4$. Clearly f is a generalized monomial, since each monomial of f has the same fingerprint X_1; however f can be written either of two ways as a sum of monomials, for $f = (w_1+w_2)X_1w_3 + w_2X_1w_4 = w_1X_1w_3 + w_2X_1(w_3+w_4)$. In general, f will not be a monomial. Of course, if W = Z, then generalized polynomials become classical polynomials, and generalized monomials become classical monomials.

For any W-algie R, and for any $r_1, r_2, \ldots$ in R, there is a unique homomorphism $W\{X\} \to R$ sending $X_i \to r_i$, for each i; we call $f(r_1, \ldots, r_m)$, the image of $f(X_1, \ldots, X_m)$ under this map, an evaluation of f in R. Let R(f) = {evaluations of f in R}; f is a GI (generalized identity) of R if R(f) = 0. Motivated by remark 1.3, given a generalized polynomial $f = \sum_{i=1}^{u} h_i$, where h_i are the generalized monomials of r, we say that f is R-proper if $R(h_i) \neq 0$ for some i (i.e. some h_i is not a GI of R); f is strongly R-proper if $\mathrm{Ann}(\sum_{i=1}^{u} R(h_i)) = 0$, and f is R-strong if f is R-proper, for every nonzero homomorphic image $\bar{R}$ of R.

Define $G_f(R) = \cup\{R(f_\pi) \mid$ all generalized monomials f_π of f}, and G(R) = the additive subgroup of R, generated by $\{G_f(R) \mid$ every GI f of R}. It is easy to see that G(R) is an ideal of R. Call R a GPI-algie if Ann G(R) = 0.

Remark 2.1. A generalized polynomial f is R-proper (resp. strongly R-proper) iff $G_f(R) \neq 0$ (resp. Ann $G_f(R) = 0$.) R has an R-proper GI iff $G(R) \neq 0$; R has an R-strong GI iff $1 \in G(R)$.

It will turn out that an improper GI will often yield intrinsic information about the coefficients of the polynomial, whereas a proper GI of R yields substantial information about R itself. For example, if a constant w in W is a GI of R, then $w \cdot 1 = 0$; if R is W-faithful, then w = 0. If $zX_1 - X_1z$ is a GI of R, then $z \cdot 1 \in Z(R)$. On the other hand, if $X_1X_2 - X_2X_1$ is a GI of R, then R is commutative. Note that a nonprime ring W will satisfy the improper GI $w_1X_1w_2$, for suitable nonzero w_1 and w_2.

If $f(X_1,\ldots,X_m)$ is a GI of R, then $f(0,\ldots,0) = 0$. Hence, f has the form $f_0 + w$, where $w\cdot 1 = 0$, and f_0 is a GI of R, having constant term 0; since $G_f(R) = G_{f_0}(R)$, it follows that our study of the structure of R will not be affected at all if we consider only GI's having constant term 0. Thus, without further ado, we shall only consider generalized monomials with constant term 0.

Define the *degree of* X_i *in* a generalized monomial h to be the degree of X_i in the fingerprint of h; h is *linear in* X_i if X_i has degree 1 in h. The *degree of* h is the sum of the degrees of each indeterminate in h, and the *degree of* a generalized polynomial f (written deg f) is the largest degree of its generalized monomials. A generalized polynomial f is *homogeneous in* X_i if X_i has the same degree in each of its generalized monomials, and f is *weakly homogeneous* if each of its generalized monomials has the same degree; f is *multilinear* if every indeterminate of f has degree 1 in each of the generalized monomials of f. Also, an element of W{X} has *normal form* if it is W-proper and multilinear of degree 2.

We can apply the notion of "evaluation" in W{X} itself, by substituting elements of W for indeterminates. A polynomial so obtained is called a *reduction*. Clearly if f' is a reduction of f, then deg f' < deg f, f' is a sum of at most as many monomials as f, and, for any W-algie R, $R(f') \subseteq R(f)$. Given a GI f of R, an R-*component* of f is a sum of generalized monomials of f which is a GI of R.

Proposition 2.2. If f is a multilinear GI of W, then either f is W-improper, or there exists a reduction of a W-component of f in normal form.

Proof. Easy (cf. [27, theorem 1].)

In view of Proposition 2.2 and other results, we shall be interested in constructing multilinear identities from arbitrary identities. This is done in two steps, as in PI-theory. A generalized polynomial $f(X_1,\dots,X_m)$ is <u>blended</u> if $f(X_1,\dots,X_{i-1},0,X_{i+1},\dots,X_m) = 0$ for each i. An easy induction argument (in [27, §1]) shows that any generalized polynomial f can be written as a sum of blended generalized polynomials f_i, such that f_i is an R-component of f, for every algie R of which f is a GI. Thus, our problem is to multilinearize blended generalized polynomials. This is a standard procedure. Given $f(X_1,\dots,X_m)$, define $\Delta_i f(X_1,\dots,X_{m+1}) =$

$$f(X_1,\dots,X_i+X_{m+1},\dots,X_m)-f(X_1,\dots,X_i,\dots,X_m)-f(X_1,\dots,X_{m+1},\dots,X_m).$$

Inductively, define $\Delta_i^{(1)}f = \Delta_i f$, and $\Delta_i^{(k)}f = \Delta_i\Delta_i^{(k-1)}f$ for $k > 1$. For any W-algie R, clearly $R(\Delta_i^{(k)}f) \subseteq R(f)$, so this procedure sends a GI of R to a GI of R.

<u>Proposition 2.3</u>. Suppose $f(X_1,\dots,X_m)$ is blended. Let h be a monomial of f, having maximal degree, and let d_i be the degree of h in X_i, for $1 \le i \le m$. Then $\Delta_i^{(d_1-1)} \dots \Delta_m^{(d_m-1)}f$ is linear in $\sum_{i=1}^m d_i$ indeterminates. If h' is a monomial of f, with degree d_i' in X_i, $1 \le i \le m$, then $\hat{h}' = \Delta_1^{(d_1-1)} \dots \Delta_m^{(d_m-1)}h' = 0$ unless $d_i' = d_i$ for each i, in which case $\hat{h}'$ is a sum of $\Pi_{i=1}^m d_i'$ monomials, each of which is sent to h' under the map $X_j \to X_i$, $i + \sum_{u=i+1}^m d_u < j < i + \sum_{u=i}^m d_u$ (for $i = 1,\dots,m$).

<u>Proof</u>. Given in [27, proposition 1].

The process of proposition 2.3 is called <u>multilinearization</u>; if g is a multilinear generalized polynomial obtained by applying various Δ_i to f, we say g is a <u>multilinearization</u> of f. An immediate consequence of proposition 2.3 is

Corollary 2.4. For any weakly homogenous, generalized polynomial f, $G_f(R) \subseteq$ additive subgroup of R generated by $\cup\{G_g(R) \mid$ multilinearizations g of f$\}$.

If R is a W-algie and H is a commutative Z(R)-algebra, then $R' = R \otimes H$ is a W-algie, the tensor product taken over Z(R). (Indeed, we define $w(\Sigma r_i \otimes h_i) = \Sigma wr_i \otimes h_i$ and $(\Sigma r_i \otimes h_i)w = \Sigma r_i w \otimes h_i$). An important example is when $H = (Z(R))[\lambda]$; then $R' = R[\lambda]$. A natural question is, "Which GI's of R are GI's of R'?" This question has been studied in depth in greater generality in [30,§2], but we remark here that every multilinear GI of R is a GI of R', and, if R is torsion-free over an infinite subring of Z(R), then every GI of R is a GI of R'. The proofs parallel the usual PI-theoretic arguments. In view of these facts, one is led to study algies in terms of the GI that they satisfy; this will be the philosophy behind §4. Accordingly, say two algies R and R' are equivalent (resp. m-equivalent) if the GI's (resp. multilinear GI's) of R and R' are the same. Clearly equivalence implies m-equivalence.

Remark 2.5. Suppose R and R' are m-equivalent, and f is a multilinear generalized polynomial. Then f is R-proper iff f is R'-proper.

Proof. If f is R-proper then some generalized monomial f_π of f is not a GI of R; hence f_π is not a GI of R', so f is R'-proper. The converse is proved symmetrically.

In the course of proofs, one occasionally wants to change the base ring of an algie. The definition of algie presented here enables us to observe that, for any W-algie R, the canonical homomorphism

$W \to W\cdot 1 \subseteq R$ induces a canonical homomorphism $W\{X\} \to R\{X\}$, which is an injection iff R is W-faithful.

§3. Generalized Identities of Endomorphism Rings

Just as the PI-theory develops around the study of identities of matrix algebras over fields, our treatment of GI-theory starts with the study of endomorphisms of vector spaces over fields. Let F be a field and let M be an F-vector space. In this section, we let $P = \mathrm{End}_F M$, and assume that P is a W-algie. Recall that the <u>socle</u> of a semiprime ring R (written soc (R)) is defined to be 0 unless R has minimal nonzero right ideals, in which case socR = Σ(minimal nonzero right ideals of R) which, by [13, p. 65, theorem 1], is Σ(minimal nonzero left ideals of R). We shall use here the characterization in [13, p. 75] of soc P as {finite-ranked transformations of $\mathrm{End}_F M$ (=P)}. If xPy is finite-ranked then x or y is finite-ranked, so we record the fact that socP is a prime ideal of P.

We shall examine socP by focusing on a particular finite-dimensional subspace of M, which we call V_0. Given $f(X_1,\dots,X_m)$ in $W\{X\}$, and numbers $u_1,\dots,u_m$, say f is $(V_0,(u_i))$-<u>valued</u> if there exist subspaces $V_1,\dots,V_m$ of M, having respective dimension $u_1,\dots,u_m$, such that $f(x_1,\dots,x_m)M \subseteq V_0$ for all x_i in P such that $x_iV_i = 0$ for $1 \le i \le m$. If f is a sum of v monomials, define ht $f = (1 + \deg f)v$. The key to the section is the following computational result:

<u>Theorem 3.1</u>. With notation as above, suppose V is u_0-dimensional, and suppose $f(X_1,\dots,X_m)$ is a multilinear generalized polynomial which is $(V,(u_i))$-valued for some $u_1,\dots,u_m$. Let $h = \mathrm{ht}(f)$,

$u = \max(u_0, u_1, \ldots, u_m)$, and let U be the additive F-subspace of P generated by F and the coefficients of f. There are functions $\psi_u : \mathbb{Z}^+ \to \mathbb{Z}^+$ with the following property: There is a generalized polynomial g with coefficients in U, equal to f in $P\{X\}$, such that each monomial of g has at least one coefficient of finite rank $\le \psi_u(h)$.

Proof. Define $\psi_u(h)$ inductively as follows: $\psi_u(0) = 0$ for all u, and $\psi_u(h) = \max(u, \psi_{u+h}(h-1), \psi_{u'}(h')$ for all $u' < u$ and all $h' \le h)$. Proceed by induction on h. Let $V_1, \ldots, V_m$ be subspaces of respective dimensions $u_1, \ldots, u_m$, with $f(x_1, \ldots, x_m)M \subseteq V$ for all x_i such that $x_i V_i = 0$, $1 \le i \le m$. Write $f = \sum_{i=1}^{m} \sum_{j=1}^{v_i} f_{ij}(X_1, \ldots, X_{i-1}, X_{i+1}, \ldots, X_m) X_i w_{ij}$ such that, if f_{ij} is a sum of v_{ij} monomials, then $\Sigma v_{ij} = v$. We may assume that $w_{ij} M \subseteq V_i$, $t_i < j < v_i$, t_i minimal within the above context. Let $f' = \Sigma_{i=1}^{m} \Sigma_{j=1}^{t_i} f_{ij} X_i w_{ij}$, and $f'' = f - f'$. Clearly f'' is $(0, (u_i))$-valued, so f' is $(V, (u_i))$-valued. If $f' \ne 0$ and $f'' \ne 0$, then $\text{htf}' < h$ and $\text{htf}'' < h$; by induction there exist generalized polynomials g' and g'' with coefficients in U, equal to (respectively) f' and f'' in $P\{X\}$, such that each monomial has a coefficient of rank $\le \psi_u(\text{htf}')$, $\psi_u(\text{htf}'')$ respectively. Since both numbers $\le \psi_u(h)$, we are done in this case, with $g = g' + g''$. If $f' = 0$, we can take $g = f$ since (by assumption) each w_{ij} has rank $\le u$. So we are done unless $f'' = 0$, i.e. $t_i = v_i$, for all i.

Choose y in M, such that $w_{11} y \in V_1$. Let $\{y_t\}$ be a basis of M. For each t, there exist x_{1t} in P and $\mu_1, \ldots, \mu_{v_1}$ in F, with $\mu_1 = 1$, such that $x_{1t} V_1 = 0$ and $x_{1t} w_{ij} y = \mu_j y_t$, $1 \le j \le v_1$. For all $x_2, \ldots, x_m$ such that $x_i V_i = 0$ and $x_i w_{ij} y = 0$ for all j, $2 \le i \le m$, we get, for all t, $\Sigma_{j=1}^{v_1} \mu_j f_{1j}(x_2, \ldots, x_m) y_t = f(x_{1t}, x_2, \ldots, x_m) y \in V$.

Therefore, setting $f'_{11} = \Sigma_{j=1}^{v_1}\mu_j f_{1j}$, we see that f'_{11} is $(V,(u_2 + v_2,\ldots,u_m + v_m))$-valued; also, $ht(f'_{11}) \leq vm \leq h-1$. Hence, by induction, there exists g'_{11} with coefficients in U, equal to f'_{11} in P{X}, with each monomial having a coefficient of rank $\leq \psi_{u+h}(h-1) \leq \psi_u(h)$. Now let $f'' = f - f'_{11}X_1w_{11}$. We can write f" as $\Sigma_{i=2}^{m}\Sigma_{j=1}^{v_i} f_{ij}X_iw_{ij} + f'''$, where $f''' = \Sigma_{j=1}^{v_1} f_{1j}X_1w_{1j} - f'_{11}X_1w_{11} = \Sigma_{j=2}^{v_1} f_{1j}X_1(w_{1j} - \mu_j w_{11})$. In this manner, f" is a sum of fewer than v monomials (since the f_{11} part has been removed), so $ht(f'') \leq (m+1)(v-1) \leq h-1$. Clearly f" is $(V,(u_1,u_2+v_2,\ldots,u_m+v_m))$-valued, so there exists g" with coefficients in U, equal to f" in P{X}, with each monomial having a coefficient of rank $\leq \psi_{u+h}(h-1) \leq \psi_u(h)$. Letting $g = g'_{11} + g''$, we are done.

Consequently, with the setup of theorem 3.1, there are functions $\phi:\mathbb{Z}^+ \to \mathbb{Z}^+$ such that, for any generalized monomial f_π of f and for all $x_1,\ldots,x_m$ in P, rank $f_\pi(x_1,\ldots,x_m) \leq \phi_u(h)$. This was the original version in [28], and the improved result was inspired by a subsequent remark of Martindale who (using other methods) observed that any GI of P can be "rewritten" in P{X} so that every monomial contains an element of soc(P).

Corollary 3.2. Every GI of P is P/socP-improper.

Proof. Set V = 0 and set all the $u_i = 0$.

Theorem 3.3. If P is a faithful algie m-equivalent to W, and if f is a P-proper GI of P, then $W \cap socP \neq 0$ (identifying W as a subring of P).

Proof. By corollary 2.4, some multilinearization of f is P-proper, so assume f is multilinear. Hence f is W-proper, so we have $w_1,\ldots,w_m$ in W and a generalized monomial f_π of f, such that $0 \neq f_\pi(w_1,\ldots,w_m) \in W \cap socP$ (by corollary 3.2).

More information can be culled from theorem 3.1. Let $G'(R)$ be the additive subgroup generated by $\{G_f(R) \mid$ multilinear GI's f of $R\}$. By proposition 2.3, $G(R) = 0$ iff $G'(R) = 0$. Also, by theorem 3.1, $G'(P) \subseteq \text{soc}P$. But socP is easily seen to be the smallest nonzero ideal of P. Hence $G'(P) = 0$ or $\text{soc}P = G'(P)$.

Theorem 3.4. $W \cap G'(P) = W \cap \text{soc}P$. If $a \in W \cap \text{soc}P$ has rank t, then S_{2t+1} is an identity both of Pa and of aP. If $W \cap \text{soc}P \neq 0$, then $\text{soc}P = G'(P)$.

Proof. Clearly we are done if $W \cap \text{soc}P = 0$, so assume $W \cap \text{soc}P \neq 0$. Suppose $a \neq 0$ is as in the hypothesis. Pa/Ann a is central simple of dimension t^2, so $X_{2t+1}S_{2t}$ is an identity of Pa, implying S_{2t+1} is an identity of Pa. Symmetrically, S_{2t+1} is an identity of aP. Now $S_{2t+1}(X_1a,\ldots,X_{2t+1}a)$ is a GI of P. Hence, for any $x_1,\ldots,x_{2t+1}$ in P, $x_1a\ldots x_{2t+1}a \in G'(P)$. Since P is prime, $x_1a\ldots x_{2t+1}a \neq 0$ for suitable $x_1,\ldots,x_{2t+1}$. Thus $G'(P) \neq 0$, so $\text{soc}(P) = G'(P)$, as was observed above.

In particular, by theorem 3.3, if P satisfies a proper GI, then $G'(P) = \text{soc } P$.

Proposition 3.5. $G'(R) = G(R)$, for any algie R.

Proof. Otherwise $G'(R) \subset G(R)$ for some algie R, so some GI f of R is $R/G'(R)$-proper. But, by proposition 2.3, there is a multilinearization g of f which is $R/G'(R)$-proper. But then $G_g(R) \not\subseteq G'(R)$, contrary to the fact that g is a multilinear GI of R. Hence, we conclude that $G'(R) = G(R)$.

We close this section with a useful result on improper GI's of P, an easy version of a more general fact in [1].

Proposition 3.6. Suppose $f(X_1) = \sum_{i=1}^{u} w_{i1}X_1w_{i2}$, and $f(x) = 0$ for all x in P such that $xV = 0$. Then either $w_{12}y \in V$ for all i, all y in M, or the $w_{i1}\cdot 1$ are F-dependent.

Proof. Assume $w_{12}y \notin V$ for some y in M. Then there exist x in R and β_i in F, with $\beta_1 = 1$, such that $xV = 0$ and $xw_{i2}y = \beta_i y$ for all i. Now let $\{y_k\}$ be an F-base of M. Take $u_t(y) = y_t$, and let $x_t = u_t x$. Then $x_tV = 0$, and $t_t w_{i2}y = \beta_i y_t$; hence $\sum_i \beta_i w_{i1} y_t = 0$ for all t, implying $\sum_i \beta_i w_{i1} 1 = 0$.

Conversely, suppose $w_1, \ldots, w_t$ are F-dependent. Then $S_t(w_1X_1, w_2X_1, \ldots, w_tX_1)$ is a GI of P, since, for any x in P, $w_1x, \ldots, w_tx$ are F-dependent. In particular, if w is algebraic of degree $t-1$, then $S_t(w^tX_1, w^{t-1}X_1, \ldots, X_1)$ is a GI of P. Of course this polynomial and all of its multilinearizations are P-improper.

§4. Embeddings of Algies

Algie will denote W-algie. The object of this section is to develop a procedure of passing information from an algie to an m-equivalent algie. The strategy is simple - given an algie R', pass to a known algie R, m-equivalent to R'; obtain a structural result and translate the conclusion back to R'. Since we know much about $R = \mathrm{End}_F M$, our goal is to find some algie of this form which is m-equivalent to R'. It is convenient to let R' = W.

Remark 4.1. Suppose W is m-equivalent to an algie R. Then R is W-faithful. Identify W with $W\cdot 1$ in R. $Z(W) = W \cap Z(R)$. Also, if R is prime, then W is prime.

Proof. If $w \cdot 1 = 0$ then $wR = 0$, so wX is a GI of R, thus of W. Since W has a multiplicative unit, $w = 0$. Now $w \in Z(W)$ iff $[w,X]$ is a GI of W, iff $[w,X]$ is a GI of R, iff $w \in W \cap Z(R)$. If $w_1Ww_2 = 0$, then w_1Xw_2 is a GI of W, hence of R, so $w_1 = 0$ or $w_2 = 0$.

Thus, if W is m-equivalent to an endomorphism ring over a field, then W is prime. Conversely, suppose W is prime. We desire the best possible result, namely, that W is m-equivalent to an endomorphism ring $\mathrm{End}_F M$ containing W. Note that, for any algie R containing W, clearly W satisfies every GI of R. Hence, to show that W and R are m-equivalent, we need merely demonstrate that every multilinear GI of W is a GI of R. This is done in a succession of interesting steps, given in various degrees of generality, leading to theorems 4.7 and 4.8. Most of the techniques are due to Amitsur [1], [3], and [13], and Martindale [21].

Proposition 4.2. If W is semiprime then W can be embedded in an algie R, equivalent to W, with $\mathrm{Nil}(R) = 0$.

Proof [19, theorem 2.5]. Let $T = \prod_i W_i$, taken over infinitely many copies of W, and let $R = T/\mathrm{Nil}(T)$. Clearly T, and thus R, satisfy every GI of W, so it remains to show $W \hookrightarrow R$. Well, there is a homomorphism $\phi: W \to R$, given by $\phi(w) = \hat{w} + \mathrm{Nil}(T)$, where $\hat{w}(i) = w$, for all i; $\ker \phi = 0$ since $\mathrm{Nil}(T) = \{x \in T \mid x(i) \in \mathrm{Nil}(W)$, and $x(i) = 0$ for all but a finite number of $i\}$.

Remark 4.3. If $\mathrm{Nil}(R) = 0$ then $R[\lambda]$ is semiprimitive, by Amitsur's theorem, and is m-equivalent to R.

Primitive will mean left primitive, i.e. having a faithful, irreducible left module.

Proposition 4.4. If W is prime and is contained in a semiprimitive algie R' which is m-equivalent to W, then W is contained in a primitive algie R which is m-equivalent to W.

Proof. Let R' be a subdirect product of primitive algies R_γ, $\gamma \in \Gamma$, where each R_γ has a faithful, irreducible module M_γ. We briefly sketch and use Amitsur's application [3] of the theory of ultrafilters. Recall that a filter F of a set S is a collection of subsets of S, such that if S_1 and S_2 are in F, then $S_1 \cap S_2 \in F$, and any set containing a member of F is in F. F is proper if $\emptyset \notin F$. A proper filter F is an ultrafilter if, for every $S_1 \subseteq S$, either $S_1 \in F$ or $(S-S_1) \in F$. By Zorn's lemma every proper filter is contained in an ultrafilter.

Amitsur's filter on Γ is defined as follows: Given r in R', let r_γ denote the canonical image of r in R_γ. A subset S_1 of Γ is in F iff there exists nonzero w in W such that $\{\gamma | w_\gamma \neq 0\} \subseteq S_1$. Note that F is a proper filter, because W is prime; hence F can be embedded in an ultrafilter F'.

Given a collection of objects $\{B_\gamma | \gamma \in \Gamma\}$, we can form the ultraproduct $(\Pi B_\gamma)/F'$, which is the direct product ΠB_γ, modulo the congruence, $b \equiv b'$ iff $\{\gamma \in \Gamma | b(\gamma) = b'(\gamma)\} \in F'$. Ultraproducts serve to preserve all the logically elementary structure carried in common by each B_γ, c.f. [12]. In particular, if $R = (\Pi R_\gamma)/F'$ and $M = (\Pi M_\gamma)/F'$, then R is a ring with faithful irreducible left module M. Moreoever, each multilinear GI of W is a GI of R', hence of each R_γ; since a GI is an elementary sentence, we see that each GI of W is a GI of R. Therefore it remains only to show that $W \hookrightarrow R$. But the canonical map $w \to \hat{w}$, where $\hat{w}(\gamma) = w$ for all $\gamma \in \Gamma$, induces a homomorphism from W to

R, and the kernel is 0, immediately seen by using the definition of F'.

If R has a faithful irreducible module M, then $D = \mathrm{End}_R M$ is a division ring, and R is a dense subring of $\mathrm{End}\, M_D$, in the sense of Jacobson [13, p. 28]. Since a GI is a condition involving only a finite number of coefficients and substitutions, we easily have

Proposition 4.5. If R is an algie and is a dense subring of $\mathrm{End}\, M_D$ (viewed as an algie in the natural way), then R and $\mathrm{End}\, M_D$ are equivalent. In fact, if $R = \mathrm{End}\, M_D$ and f is a generalized polynomial such that $f(x_1,\ldots,x_m) = 0$ for all x_i in socR, then f is a GI of R.

Also, with this notation, let F be a maximal subfield of D; viewing $\mathrm{End}_R M \subseteq \mathrm{End}_{\mathbb{Z}} M$ and $\mathrm{End}\, M_D \subseteq \mathrm{End}_{\mathbb{Z}} M$, let R_F denote the subring of $\mathrm{End}_{\mathbb{Z}} M$ generated by F and R. M is a faithful, irreducible R_F-module, under the action $(\Sigma r_i\alpha_i)y = \Sigma(r_i y)\alpha_i$, for r_i in R, α_i in F, and y in M.

Remark 4.6. With the above notation, R and R_F are m-equivalent.

It is easy to see that $F = \mathrm{End}_{R_F} M$. Hence, letting $P = \mathrm{End}_F M$, we see that R_F and P are equivalent, by proposition 4.5. Putting 4.2-4.6 together, we have injections of a prime ring W $\subseteq$ ring with 0 nilradical $\subseteq$ semiprimitive ring $\subseteq$ primitive ring $\subseteq$ dense subring of $\mathrm{End}_F M \subseteq \mathrm{End}_F M$, with equivalence or m-equivalence at each step. Therefore we have the following major embedding theorem:

Theorem 4.7. If W is a prime ring then there exists an algie $\mathrm{End}_F M$ containing W, which is m-equivalent to W (where F is a suitable field and M is a suitable F-vector space).

Starting instead with a semiprime ring W, we get a parallel result whose proof does not require proposition 4.4.

Theorem 4.8. If W is a semiprime ring then there exist fields F_γ and F_γ-vector spaces M_γ, such that $W \subseteq \prod_\gamma \text{End}_{F_\gamma} M_\gamma$, which is m-equivalent to W.

Proof. Since a semiprimitive ring is a subdirect product of primitive images, it is clearly equivalent to the direct product of these images. Hence we can put together injections $W \subseteq$ ring with 0 nilradical $\subseteq$ semiprimitive ring $\subseteq$ direct product of primitive rings $\subseteq$ direct product of $\text{End}_{F_\gamma} M_\gamma$, for suitable fields F_γ and F_γ-vector spaces M_γ. Since there is equivalence or m-equivalence at each step, the theorem is proved.

§5. The Structure Theories of Algies with GI

In this section we apply the strategy stated at the beginning of §4, applying the embedding theorems of §4 to the results of §3 (specifically theorem 3.1), to obtain generalizations of theorems A-E of section 1. We have three conditions to examine: proper GI, strongly proper GI, and strong GI. It turns out that the last two conditions are not equivalent, even for primitive algies. In fact, theorem 5.2 reduces the study of strong GI's to PI-rings.

Suppose R is semiprime. By theorem 4.8, there is a set of algies $\{R_\gamma = \text{End}_{F_\gamma} M_\gamma | \gamma \in \Gamma\}$, such that $R \subseteq R' = \Pi\{R_\gamma | \gamma \in \Gamma\}$, which is m-equivalent to R. Given r in R', let r_γ denote the component of r in R_γ, and let $H(R) = \{r \in R | \{\text{rank } r_\gamma | \gamma \in \Gamma\}$ is bounded$\}$, an ideal of R.

Theorem 5.1. Suppose R is semiprime, with R' as above. Then $H(R) = \{r \in R | Rr$ satisfies a standard identity$\} = \{r \in R | Rr$ is a PI-ring (without 1)$\} = \{r \in R | rR$ satisfies a standard identity$\}$; thus $H(R)$ is independent of the choice of R'. $G'(R) \subseteq H(R)$, and $H(R)/G'(R) \subseteq \Sigma$(nilpotent ideals of $R/G'(R)$). In fact, for any x in $H(R)$, $RxR/G'(R)$ is nilpotent.

Proof. Suppose $r \in H(R)$. Then, for suitable t, $\max\{\text{rank } r_\gamma | \gamma \in \Gamma\} \leq t$. By theorem 3.4, each $R_\gamma r_\gamma$ and $r_\gamma R_\gamma$ satisfy the identity S_{2t+1}, so S_{2t+1} is an identity of $R'r$ and rR', hence of Rr and of rR.

Conversely, suppose Rr satisfies a strong classical identity f which, by proposition 2.3, we may assume is multilinear. Then $f(X_1 r,\dots,X_k r)$ is a GI of R; thus, by theorem 3.1 (viewing each R_γ as R-algie), $\text{rank}(x_{1\gamma} r_\gamma x_{2\gamma} r_\gamma \cdots x_{k\gamma} r_\gamma) \leq \phi_0(\text{ht}(f(X_1 r,\dots,X_k r)))$ for all γ, all $x_{1\gamma},\dots,x_{k\gamma}$ in R_γ. Since $\text{soc}R_\gamma$ is a prime ideal of R_γ, we have $r_\gamma \in \text{soc}R_\gamma$; hence there exists x_γ with $r_\gamma x_\gamma r_\gamma = r_\gamma$. Letting $x_{1\gamma} = 1$ and $x_{2\gamma} = \dots = x_{k\gamma} = x_\gamma$, we see that $\text{rank } r_\gamma \leq \phi_0(\text{ht}(f(X_1 r,\dots,X_k r)))$, for each γ. Hence $r \in H(R)$. This proves $H(R) = \{r \in R | Rr$ satisfies a standard identity$\}$; by symmetry, $H(R) = \{r \in R | rR$ satisfies a standard identity$\}$.

It is immediate, from the definitions and theorem 3.1, that $G'(R) \subseteq H(R)$. Finally, if $r \in H(R)$ then, for some k, $S_k(X_1 r,\dots,X_k r)$ is a GI of R, implying $(Rr)^k \subseteq G'(R)$ proving the last assertion.

Theorem 5.2. If an arbitrary algie R has an R-strong GI, then R is a PI-ring.

Proof. {algies with R-strong GI's} is closed under direct powers, so, in view of theorem 1.1, we may assume that $\text{Nil}(R) = 0$. Apply the notation of theorem 5.1. $1 \in G(R)$, so by theorem 5.1 and proposition 3.5, $1 \in H(R)$, implying $R = 1R$ satisfies some standard identity.

Theorem 5.3. A semiprime algie R is a GPI-algie iff $\mathrm{Ann}H(R) = 0$. In this case, R has no nonzero nil left or right ideals.

Proof. If R is a GPI-algie then, by definition, $\mathrm{Ann}G(R) = 0$; hence $\mathrm{Ann}H(R) = 0$, by theorem 5.1. Conversely, assume that $\mathrm{Ann}H(R) = 0$. Since R is semiprime, $G(R) \cap \mathrm{Ann}G(R) = 0$. Suppose $x \in H(R) \cap \mathrm{Ann}G(R)$. By theorem 5.1, $(RxR)^t \subseteq G(R)$ for some t. But then $(RxR)^t \subseteq G(R) \cap \mathrm{Ann}G(R) = 0$, so $x = 0$. Thus $0 = H(R) \cap \mathrm{Ann}G(R) \supseteq H(R)\mathrm{Ann}G(R)$, so $\mathrm{Ann}G(R) \subseteq \mathrm{Ann}H(R) = 0$, proving the first assertion.

Suppose L is a nonzero nil left ideal of R. Then $0 \neq H(R)L \subseteq H(R) \cap L$. By theorem 5.1, for any nonzero x in $H(R) \cap L$, Rx is a nil ring without 1, satisfying a polynomial identity. But $Rx/\mathrm{Ann}\, x$ is semiprime, as is easy to see, and is a nil PI-ring without 1. Thus, by [13, p. 232], $Rx/\mathrm{Ann}\, x = 0$, so $xRx = 0$, implying $x = 0$, contrary to the stipulation $H(R) \cap L \neq 0$. We must conclude that $L = 0$. Symmetrically, R has no nonzero nil right ideals.

Proposition 5.4. A semiprime algie R has an R-proper GI iff R has a left ideal satisfying a standard identity.

Proof. Using theorem 5.1, we see that R has an R-proper GI iff $G(R) \neq 0$, iff $H(R) \neq 0$, iff R has a left ideal satisfying a standard identity.

The conclusions of theorem 5.1, theorem 5.3, and proposition 5.4 are quite strong, for any prime ring having a large left ideal satisfying an identity is itself a PI-ring (Belluce and Jain [8]). The method of "viewing" a generalized identity as a polynomial identity of a left (or right) ideal was initiated by Jain [15].

Note that theorem 5.3 generalizes Corollary C, to GPI-algies. Theorem B cannot be generalized, however, as seen by the obvious example of $\mathrm{End}_F M$, where M is infinite-dimensional. We look now towards generalizing theorem A; i.e. we want a structural characterization of primitive algies with proper GI. The first major results along these lines are by Amitsur [1].

<u>Theorem 5.5.</u> Suppose R is a dense subring of $\mathrm{End}\, M_D$, for a suitable division ring D and right D-vector space M. Then $G(R) = G'(R) \subseteq \mathrm{soc}(R)$. If R has an R-proper GI, then $G(R) = G'(R) = \mathrm{soc}R$, $W\cdot 1 \cap \mathrm{soc}R \neq 0$, and D is a central simple PI-ring.

<u>Proof.</u> By proposition 4.5 and remark 4.6, there is a maximal subfield F of D, such that R is equivalent to $\mathrm{End}_F M$. Let $P = \mathrm{End}_F M$. Clearly every element of $R \cap \mathrm{soc}P$ is a finite ranked transformation in $\mathrm{End}\, M_D$, so $R \cap \mathrm{soc}P \subseteq \mathrm{soc}R$. Hence, by the observation after theorem 3.3, $G(R) = G'(R) \subseteq R \cap G'(P) \subseteq R \cap \mathrm{soc}P \subseteq \mathrm{soc}R$.

Now suppose that R has an R-proper GI. Then $G(R) \neq 0$. Since socR is the minimal nonzero ideal of R, we have $G(R) = \mathrm{soc}R$. Also, by corollary 3.2, $W\cdot 1 \cap \mathrm{soc}R \neq 0$. Choose $w \neq 0$ in $W\cdot 1 \cap \mathrm{soc}R$. If t is the F-rank of w, then $t = (wM:F) = (wM:D)(D:F)$, implying $(D:F) \leq t$; hence D is central simple (of dimension $\leq t^2$), and is thus a PI-ring.

A reasonably good bound can be found for the PI-degree of D, using the normal form, as in [27, theorem 2]. Theorem 5.5 has an easy converse: Suppose R is a dense subring of $\mathrm{End}\, M_D$, and $\mathrm{soc}R \neq 0$ and D is a PI-ring. Then there is an idempotent e of R, with $eRe \approx D$, and there is a classical identity $f(X_1,\ldots,X_m)$ of D; hence $f(eX_1e,\ldots,eX_me)$ is an R-proper GI of R.

Theorem 5.5 generalizes theorem A. Indeed, suppose, with notation as in theorem 5.5, that R is a PI-ring. Then $1 \in G(R) = \text{soc}R$, implying $R = \text{soc}R$. Since D is also a PI-ring, R is central simple.

We close this section by observing that theorem D is a consequence of the theory of §§3-4. Indeed, suppose that R is a prime PI-ring. Then R is m-equivalent to some ring $R' = \text{End}_F M$, which contains R. Hence R' is a PI-ring, implying R' is simple artinian. Let $Z = Z(R')$. RZ is a Z-subalgebra of R', m-equivalent to R'. It is well-known (and patent) that any Z-subalgebra of R' has smaller PI-degree; hence RZ = R'. Theorem D follows (cf. [5], for example).

§6. Applications of GI-theory to Classical Polynomials

At certain stages in this theory, we have made use of PI-theory to obtain results in the theory of generalized identities. At this juncture the process will be reversed, and we use the theory of generalized identities to obtain immediate answers to problems arising in PI-theory.

Let $[x,y]$ denote $xy-yx$, and let $C_R(r) = \{x \in R \mid [x,r] = 0\}$. The first result is essentially a theorem of Montgomery-Smith [31], formulated as in [28].

Theorem 6.1. Suppose there exists r in R, and nonzero $g(\lambda)$ in $Z(R)[\lambda]$, such that $g(r) = 0$, $C_R(r)$ is a PI-algebra, and $g'(r)$ is invertible in R, where $g'(\lambda)$ is the formal derivative of g. Then R is a PI-algebra.

Proof [28, theorem 4]. Let $g(\lambda) = \sum_{i=0}^{m} \alpha_i \lambda^i$, and define $f(X) = \sum_{k=1}^{m} \alpha_k \sum_{i=0}^{k-1} r^i X r^{k-1-i}$. Note, for all x in R, that $f(x) \in C_R(r)$, and $f(1) = f'(r)$ is invertible in R. If $h(X_1,\ldots,X_n)$ is a multilinear, strong classical identity of $C_R(r)$, then $h(f(X_1),\ldots,f(X_n))$ is a GI of R, one of whose generalized monomials is $\pm f(X_{i_1})\ldots f(X_{i_n})$; specializing each X_i to 1, we see $\pm(g'(r))^n \in G(R)$, implying $1 \in G(R)$, so R is a PI-ring, by theorem 5.2.

A related question is, "If f is a classical polynomial and if $[r,f]$ is a GI of R, for $r \notin Z(R)$, is R a PI-ring?" This question can be generalized as follows: "If R is an algie, and f is a classical polynomial and g is a generalized polynomial such that $[f,g]$ is a GI of R, under what conditions is R a PI-ring?" These questions have reasonably sharp answers, which will be given in the next sequence of results.

Remark 6.2. If a semiprime ring R is a subdirect product of R_1 and R_2, then there are semiprime homomorphic images R_i' of R_i, $i = 1,2$, such that R is a subdirect product of R_1' and R_2'.

Proof. Let $R_i = R/A_i$, and let $A_i' = \cap\{\text{prime ideals of } R \supseteq A_i\}$. Each R/A_i' is semiprime. Moreover, for any prime ideal P of R, we have $0 = A_1A_2 \subseteq P$, so $A_1 \subseteq P$ or $A_2 \subseteq P$, implying $A_1' \cap A_2' \subseteq P$. Therefore, $A_1' \cap A_2' \subseteq \cap\{\text{prime ideals of } R\} = 0$, so R is a subdirect product of R/A_1' and R/A_2'.

Proposition 6.3. Suppose $g(X_1,\ldots,X_m)$ is a generalized polynomial in $W\{X\}$, and R is an algie of the form $\mathrm{End}_F M$, F a field and M an F-vector space, such that $R(g) \subseteq F$. Then either $(g-g(0,\ldots,0))$ is a GI of R or R is central simple of degree $\leq$ some number determined only by the properties of g.

Proof. Subtracting $g(0,\ldots,0)\cdot 1$, we may assume that the constant term of g is 0. By proposition 4.5, $g(x_1,\ldots,x_m) \neq 0$ for suitable $x_1,\ldots,x_m$ in socR. But then $g(x_1,\ldots,x_m) \in F \cap \text{socR}$, implying $1 \in \text{socR}$. Thus R is central simple.

To obtain the bound on the degree, we apply the Δ_i step (described before proposition 2.3) repeatedly, for various i, in order to obtain a generalized polynomial g' with $0 \neq R(g') \subseteq F$, such that $g'(X_1,\ldots,X_t)$ is additive in the sense that $g'(\ldots,x_{i1}+x_{i2},\ldots) = g'(\ldots,x_{i1},\ldots) + g'(\ldots,x_{i2},\ldots)$ for all i, all x_{i1},x_{i2} in R. (The formal proof for this kind of assertion is given in [25, §1], but the reason is clear - we can apply some Δ_i unless the generalized polynomial is additive, so we continue until we obtain either an additive generalized polynomial or a multilinear one, which is automatically additive.) Suppose g has degree d, and g can be written as a sum of v monomials. Clearly we can only apply the Δ step at most $d-m$ times, at each stage multiplying the number of monomials by $\leq 2^{d-m}$. Thus, setting $k = 2^{(d-m)(d-m)}$, we see that g' can be written as a sum of kv monomials. Since g' is additive, there exist $x_1,\ldots,x_t$ having rank 1, such that $0 \neq g'(x_1,\ldots,x_t) \in F$. But this element has rank $\leq kv$; hence M has dimension $\leq kv$, implying R has degree $\leq kv$.

Theorem 6.4. Suppose R is a semiprime algie, f is a strong classical polynomial of degree d, and g is a generalized polynomial of degree m, such that $[f,g]$ is a GI of R. Then R is the subdirect product of a PI-ring R_0 and an algies R_1, such that g is the sum of a GI of R_1 and a constant $(g(0,\ldots,0))$ whose image in R_1 lies in $Z(R_1)$. If R is prime then R equals R_0 or R_1.

Proof. Embed R in $\{R_\gamma \mid \gamma \in \Gamma\}$ as in theorem 4.8, where each R_γ has the form $\text{End}_{F_\gamma} M_\gamma$ and satisfies every multilinear GI of R. (If R is prime, then we may assume that Γ has one element, by theorem 4.7.) Specifying one particular γ in Γ, we write $M = M_\gamma$ and $F = F_\gamma$; it suffices to prove either that R_γ has bounded PI-degree (bounded by a function of d and the properties of g) or that g is the sum of a GI of R_γ and a constant ($g(0,\ldots,0)$) whose canonical image in R_γ lies in F. Given an element r in R, let $\bar{r}$ denote the canonical image in R_γ.

Clearly we are done unless $\dim_F M > [d/2]$. Let $\{y_k\}$ be an F-base of M, and define e_{ij} by $e_{ij}(y_j) = y_i$ and $e_{ij}(y_k) = 0$ for $j \neq k$. Now, multilinearizing, we obtain an R_γ-proper, multilinear classical polynomial f' of degree $\leq d$, such that $[f',g]$ is a GI of R. Assume that $X_1 \ldots X_d$ has nonzero coefficient u in f'. For any element a in $R(g)$, $[f',a]$ is a multilinear GI of R, and is thus a GI of R_γ. But $f(e_{11},e_{12},e_{22},\ldots) = ue_{1t}$; by symmetry, each $ue_{ij} \in R_\gamma(f)$. Hence, $u[e_{ij},\bar{a}] = [ue_{ij},\bar{a}] = 0$ for each i,j, implying $[\operatorname{soc} R_\gamma,\bar{a}] - 0$. By the density theorem, $\bar{a} \in F$. Therefore $R_\gamma(g) \subseteq F$. In particular, $g(0,\ldots,0) \in F$, and the theorem is proved by proposition 6.3.

Note, by remark 6.2, that we may conclude that R_0 and R_1 in theorem 6.4 are both semiprime. Also, setting $f = X_1$, we see easily that "generalized central polynomials" do not play a role in the GI structure theory, since there existence implies a PI-condition. The last application of GI-theory to PI-theory in this section will be via the ideal $H(R)$, defined before theorem 5.1. Say a subset T of R is PI if, for some strong classical polynomial $f(X_1,\ldots,X_m)$, $f(x_1,\ldots,x_m) = 0$ for all x_i in T.

<u>Remark 6.5</u>. Suppose A is a left ideal of R and let A' be the subring of R generated by A and $Z(R)$. A is PI iff A' is a PI-ring. (Proof: $[a_1+z_1,a_2+z_2] = [a_1,a_2] \in A$, for a_i in A, z_i in $Z(R)$. Hence,

if $f(X_1,\ldots,X_m)$ is a strong classical identity of A, then $f([X_1,X_2],\ldots,[X_{2m-1},X_{2m}])$ is a strong classical identity of A'.)

We say an element a of R has PI-rank t if S_{2t+1}, but not S_{2t-1}, is an identity of Ra. This is the same as saying $S_{2t+1}(X_1a,\ldots,X_{2t+1}a)$ is a GI of R, so the PI-rank of a does not change if we replace R by an m-equivalent algie containing R.

Lemma 6.6. Suppose $R = \mathrm{End}_F M$, M a vector space over a field F, and $a \in R$. The following sentences are equivalent:

(i) Ra satisfies a strong classical identity of degree $\leq 2t+1$.

(ii) a has rank $\leq t$.

(iii) Ra is a sum of $\leq t$ minimal nonzero left ideals of R.

(iv) a has PI-rank $\leq t$.

Proof. Let $\{y_k\}$ be an F-base of M, and define e_{ij} by $e_{ij}(y_j) = y_i$ and $e_{ij}(y_k) = 0$ if $j \neq k$.

(i) $\rightarrow$ (iii). By theorem 5.1, $a \in \mathrm{soc}R$. Hence Ra is a sum of u minimal nonzero ideals of R, and we are done unless $u > t$. Since each minimal nonzero left ideal contains an idempotent, we may assume that $Ra = Re_{11} + Re_{22} + \ldots + Re_{uu}$. But then, if $f(X_1,\ldots,X_{2t+1})$ is a multilinear strong classical identity of Ra, we have $0 = f(e_{11},e_{12},e_{22},\ldots,e_{t+1,t+1}) = e_{11}e_{12}\cdots e_{t+1,t+1} = e_{1,t+1}$, a contradiction. Hence $u \leq t$ after all.

(ii) $\leftrightarrow$ (iii) is well-known.

(ii) $\rightarrow$ (iv). Done in theorem 3.4.

(iv) $\rightarrow$ (i). Trivial.

Since condition (ii) in lemma 6.6 is left-right symmetric, it follows that a has PI-rank t iff S_{2t+1}, but not S_{2t-1}, is an identity of aR. Also note that, by theorem 3.1, a has PI-rank iff Ra is a

PI-ring (without 1) iff R has a proper GI with every coefficient in F(a).

Proposition 6.7. If R is semiprime and a_1, a_2 have respective PI-ranks t_1 and t_2 in R, then a_1+a_2 has PI-rank $\leq t_1+t_2$.

Proof. Let $R' = \Pi P_\gamma$, where each $P_\gamma = \text{End}_{F_\gamma} M_\gamma$, as in theorem 4.8, with R' m-equivalent to R. Given r in R', let r_γ denote the image of r in P_γ. By lemma 6.6, $(a_1)_\gamma$ and $(a_2)_\gamma$ have respective ranks t_1 and t_2 for each γ, so $(a_1+a_2)_\gamma$ has rank $\leq t_1+t_2$; hence each $(a_1+a_2)_\gamma$ has PI-rank $\leq t_1+t_2$. Therefore a_1+a_2 clearly has PI-rank $\leq t_1+t_2$.

Given a subset T of R, define PI-rank T = max{PI-rank of x| x ε T}. Also, let $T_1+T_2 = \{x_1+x_2 | x_1 \in T_1, x_2 \in T_2\}$.

Corollary 6.8. If R is semiprime, then PI-rank $(T_1+T_2) \leq$ PI-rank T_1 + PI-rank T_2.

Theorem 6.9. For any R and any left ideals A_1 and A_2 of R, if A_1 and A_2 are PI then A_1+A_2 is PI.

Proof. First assume that R is semiprime. By lemma 6.6, A_1 and A_2 each have finite PI-rank, so, by corollary 6.8, A_1+A_2 has finite PI-rank. In other words, for some k, $S_{2k+1}(X_1X_{2k+2}, \ldots, X_{2k+1}X_{2k+2})$ is an identity of A_1+A_2, which is therefore PI.

In general, we may replace R by the subring generated by A_1, A_2, and Z(R); then we need to show that R is a PI-ring. We have just seen that this is true if R is semiprime, and the assertion follows in general from theorem 1.1.

The degrees of the identities involved in theorem 6.9 can be examined with some care, as in [28]. Note that theorem 6.9 implies that the sum of all PI-left ideals of R is an ideal (not necessarily PI); if R is semiprime then this ideal is clearly H(R).

There remains some investigation to be done on the important ideal H(R), which is not too well behaved. (Indeed, from the definition, we see that $H(\Pi R_i) \neq \Pi H(R_i)$, in general.) For example, let $H'(R) = \{x \in R | Rx \text{ is PI}\}$. H'(R) = H(R) when R is semiprime. When does H'(R/H'(R)) = 0? We could define inductively $H_0(R) = H'(R)$ and $H_\alpha(R) = \cup\{H_\beta(R) | \beta < \alpha\}$ for ordinals α. Under what conditions on R does $H_\alpha(R) = R$ for some α? For such rings R we could build a "GI-induction" procedure.

§7. The Martindale Ring of Quotients

One of the major early results of the theory of generalized polynomial identities was Martindale's use of the "central closure" of a prime ring, related to the maximal quotient ring, to show that a prime ring with suitable generalized polynomial identity has a quotient ring which is primitive with nonzero socle (cf. [19]). Later, Martindale [21] developed an embedding procedure (inspiring the theorems here in §4), with which he derived his theorem from Amitsur's central closure. We shall use that method here, and shall use recent PI-theory to obtain a sharper focus on the central closure of prime algies with proper GI. The reader may be interested to read Amitsur's generalization of the central closure for arbitrary semiprime rings, in [5]; at the end of this section there will be a discussion of the maximal quotient ring of a semiprime PI-algie.

The following construction works even without the assumption that R has a unit element. Until further notice, suppose R is prime, and let $\mathcal{A}$ = {Ideals of R}. Viewing each member of $\mathcal{A}$ as a <u>left</u> ideal, define $Q_0(R) = \varinjlim \mathrm{Hom}(A,R)$, taken over the filter $\mathcal{A}$. In other words, let (f,A) denote a left module homomorphism $f:A \to R$, for A in $\mathcal{A}$, and define an equivalence on the pairs (f,A), by $(f_1,A_1) \sim (f_2,A_2)$ if there exists $A \subseteq A_1 \cap A_2$ such that $f_1|_A = f_2|_A$. Denoting the equivalence class of (f,A) by [f,A], we let $Q_0(R) = \{[f,A] \mid A \in \mathcal{A},\ f \in \mathrm{Hom}(A,R)\}$, endowed with ring structure $[f_1,A_1]+[f_2,A_2] = [f_1+f_2, A_1 \cap A_2]$ and $[f_1,A_1][f_2,A_2] = [f_1 \circ f_2, A_2A_1]$. Let $Z = Z(Q_0(R))$. Z is called the <u>extended centroid of R</u>. Z is a field; also $[f,A] \in Z$ iff f is a <u>bimodule</u> homomorphism; hence the definition of Z is left-right symmetric. Let $\hat{R}$ be the subring $RZ \subseteq Q_0(R)$. $\hat{R}$ is called the <u>central closure</u> of R. Clearly $\hat{R}$ is m-equivalent to R. Also, for any x in $\hat{R}$, one sees easily that $R \cap Rx \neq 0$ and, symmetrically, $R \cap xR \neq 0$. Also, $Q_0(Q_0(R)) = Q_0(R)$, implying $\hat{\hat{R}} = \hat{R}$. These facts are all in [15], and the proofs are straightforward. Another result, also due to Martindale [19], is the following theorem, whose easy proof given here is inspired by an idea in [9, theorem 3.1].

<u>Theorem 7.1</u>. Suppose $R' \supseteq \hat{R}$, and R' is an algebra over a field Z' containing Z. Then $\hat{R} \otimes_Z Z' \approx \hat{R}Z'$ (viewed in R').

<u>Proof</u>. Otherwise we have Z-independent elements $\alpha_1,\ldots,\alpha_t$ in Z' and nonzero elements $r_1,\ldots,r_t$ in $\hat{R}$, such that $\sum_{i=1}^{t} r_i\alpha_i = 0$; choose such a situation with t minimal. Clearly $t > 1$. Define a map $f:\hat{R}r_t\hat{R} \to \hat{R}r_1\hat{R}$, by $f(\sum_j r_{1j}r_tr_{2j}) = \sum r_{1j}r_1r_{2j}$. If $\sum_j r_{1j}r_tr_{2j} = 0$, then, letting $r_i' = \sum_j r_{1j}r_ir_{2j}$, $1 \le i \le t$, we see that $\sum_{i=1}^{t-1} r_i'\alpha_i = 0$; by the minimality of t, each $r_i' = 0$, so $\sum r_{ij}r_1r_{2j} = 0$. Hence f is well-

defined, and is clearly an R-bimodule homomorphism. Let $z = [f, \hat{R}r_t\hat{R}] \in Z$. Then $r_1 = r_t z$, so $r_1(\alpha_1 + z^{-1}\alpha_t) + r_2\alpha_2 + \ldots + r_{t-1}\alpha_{t-1} = 0$, contrary to the minimality of t.

At this point we focus on a procedure used to simplify some of the structure theory. Suppose r is an element of R. Clearly $\mathrm{Ann}_{Rr} r$ is an ideal of Rr; let $\overline{Rr}$ denote $Rr/\mathrm{Ann}\ r$; ($\overline{}$ shall always denote canonical homomorphic images).

Remark 7.2. If R is prime (resp. semiprime) then $\overline{Rr}$ is prime (resp. semiprime). If R has faithful, irreducible module M, then $M/\{y \in M | ry = 0\}$ is a faithful, irreducible $\overline{Rr}$-module.

Proof. Suppose $\overline{ar}\ \overline{Rr}\ \overline{br} = 0$. Then $(rar)R(rbr) = 0$. If R is prime (resp. if R is semiprime and $a = b$) then $rar = 0$, so $\overline{ar} = 0$, proving the first assertion. To see the second assertion, let $M' = \{y \in M | ry = 0\}$. If $arM \subseteq M'$, then $rarM = 0$, so $rar = 0$, implying $\overline{ar} = 0$; hence M/M' is $\overline{Rr}$-faithful. Likewise, if M''/M' is a nonzero submodule of M/M', then RrM'' is a nonzero submodule of M, implying $RrM'' = M$. Thus $M/M' = \overline{Rr}M''/M' \subseteq M''/M'$, so M/M' is $\overline{Rr}$-irreducible.

If A is a subring of R, then $(A+Rr)/\mathrm{Ann}\ r \approx A/(A \cap \mathrm{Ann}\ r)$; thus, if $r \in A$, there is no ambiguity involved in writing $\overline{Ar}$. Of course, the situation with respect to rR is completely symmetrical. Note that $\overline{Rr}$ need not have 1.

Call R strongly primitive if R is dense in a ring of the form $\mathrm{End}\, M_D$, D a PI-division ring and M a right D-vector space. In this situation, for any r in soc R, $\overline{Rr}$ satisfies a classical identity (cf. theorem 5.5) and thus, by remark 7.2 and Kaplansky's theorem, $\overline{Rr}$ is central simple. This implies that $\overline{Rr}$ has a unit element, which we shall temporarily call $\overline{r_1 r}$. It follows that $(r_1 r)^2$ is an idempotent, whose image in $\overline{Rr}$ is the unit element $\overline{r_1r}$. Moreover, if $F = Z(D)$ then socR is an F-algebra, and $Rr \subseteq$ socR can also be viewed canonically as F-algebra. Since $F \cdot \text{Ann } r \subseteq \text{Ann } r$, this algebra structure passes to $\overline{Rr}$. Suppose $\overline{xr} \in Z(\overline{Rr})$. Then $rxrX_1r - rX_1rxr$ is a GI of R, hence of $\text{End}_F M$ by proposition 4.5 and remark 4.6. Therefore, by proposition 3.6, $rxr = \alpha r$, for some α in F, implying $\overline{xr}\;\overline{r_1r} = \overline{r_1r}$. This proves (with $e = (r_1r)^2$)

Lemma 7.3. Suppose R is strongly primitive and let $F = Z(D)$, D as above. Then, for any r in socR, $\overline{Rr}$ is a central simple (finite dimensional) $\overline{Fe}$-algebra, for a suitable idempotent e of Rr.

(We continue to assume that R is prime.)

Remark 7.4. If $\overline{Rr}$ is simple artinian, then R is primitive and $r \in$ socR. If $\overline{Rr}$ is also a PI-ring, then R is strongly primitive.

Proof. Rr must be a sum of minimal nonzero left ideals of R, any of which is a faithful irreducible R-module. Write R dense in $\text{End}\, M_D$. $\overline{Rr}$ has a canonical image of D; thus, if $\overline{Rr}$ is a PI-ring, D must certainly be a PI-ring.

These results are all well-known, but it is interesting to see how the ideas are illuminated by GI-theory. The groundwork is laid for this section's major result.

Theorem 7.5. If R satisfies a proper GI, then $\hat{R}$ is strongly primitive. In fact, $G(R) \subseteq \text{soc}\hat{R}$.

Proof. Using theorem 4.7, embed $\hat{R}$ in some algie $P = \text{End}_F M$ which is m-equivalent to R. By theorem 3.4, $\text{soc}P = G(P) \supseteq G(R) \neq 0$. By lemma 7.3, for any r in $G(R) \cap \text{soc}P$, $\overline{Pr}$ is central simple of some dimension t^2, and has multiplicative unit $\bar{e}$ for some idempotent e of Pr.

Clearly $Z \subseteq F$, and the action of F on $\overline{Pr}$ induces an action of Z on $\overline{\hat{R}r}$. Elements $\overline{r_1 r}, \ldots, \overline{r_k r}$ are Z-independent (for r_i in R) iff $rr_1 r, \ldots, rr_k r$ are Z-independent, iff $\overline{rr_1 r}, \ldots, \overline{rr_k r}$ are F-independent (by theorem 7.1), implying $k \leq t^2$. Thus $\overline{\hat{R}r}$ is a prime ring which is finite-dimensional over Z, and must therefore be central simple. Hence, by remark 7.4, $\hat{R}$ is strongly primitive and, by theorem 5.5, $G(R) \subseteq \mathrm{soc}\hat{R}$.

This theorem generalizes slightly [21, theorem 2.10]. We can obtain more information, however, by looking carefully at $\hat{R}$. As in theorem 7.5, suppose $\hat{R}$ is a dense subring of $\mathrm{End}\, M_D$, D a PI-division ring and M a right D-vector space. Let $\{y_k\}$ be a D-base of M, and let e_{ij} send y_i to y_j, and all other y_k to 0. Each $e_{ij} \in \mathrm{soc}\hat{R}$. Now let B be an ideal of R. Let $A_i = \{r \in R | e_{ii} r \in R\}$ and $A'_j = \{r \in R | re_{jj} \in R\}$. A_i (resp. A'_j) is a nonzero right (resp. left) ideal of R, so $0 \neq e_{ii} A_i B A'_j e_{jj} \subseteq B \cap e_{ii} R e_{jj} = B \cap e_{ij} D$. Thus, for any e_{ij}, we have d_{ij} in D, such that $d_{ij} e_{ij} \in B$. This kind of fact has been used by Smith [31].

<u>Proposition 7.6</u>. Suppose e is an idempotent in $\mathrm{soc}\hat{R}$. Then $e\hat{R}e \approx \overline{\hat{R}e}$, as Z-algebras, and $e\hat{R}e$ is the ring of central quotients of eRe.

<u>Proof</u>. The isomorphism is given by $exe \to \overline{xe}$, for any x in $\hat{R}$, and is well-known. To prove the other assertion, let $T = eRe \subseteq e\hat{R}e$. Any element of $e\hat{R}e$ has the form $e(\Sigma\alpha_i r_i)e = \Sigma\alpha_i(er_i e)$, for suitable α_i in Z, r_i in R. Now if α_i is represented by the bimodule homomorphism $f{:}B \to R$, then $f{:}eBe \to T$ is a T-bimodule homomorphism, and eBe is an ideal of T. Thus, we can view $e\hat{R}e \subseteq \hat{T}$. But $e\hat{R}e$ is simple artinian, so $Q_0(e\hat{R}e) = e\hat{R}e$, implying $\hat{T} \subseteq e\hat{R}e$. Hence $\hat{T} = e\hat{R}e$. By [19, p. 583],

$e\hat{R}e$ is the classical left and right ring of quotients of T; since $e\hat{R}e$ is a PI-ring, $e\hat{R}e$ is the ring of central quotients of T.

Continuing the above notation, observe that $Q_0(\hat{R}) = \text{End}\, M_D$, so $Z = Z(Q_0(R)) = Z(Q_0(\hat{R})) = Z(I)$. Thus, although D depends on the choice of M, Z(D) is uniquely determined. (In fact, D itself is determined up to isomorphism, cf. [13, pp. 44-45]. By remark 7.3 and proposition 7.6, we also have $Z(e\hat{R}e) = Ze$. In particular, Z(eRe) is an integral domain whose field of fractions is Ze.

Theorem 7.7. Suppose R is a proper algie with GI, with the above notation, and let B be an ideal of R. (i) For any x in $\text{soc}\hat{R}$, there exists nonzero α in $Z \cap B$ with $\alpha x \in B$. (ii) For any r in $R \cap \text{soc}\hat{R}$, $\overline{\hat{R}r}$ is the ring of central quotients of $\overline{Rr}$.

Proof. (i) Any element x in $\text{soc}\hat{R}$ has the form $\sum_{i,j=1}^{k} e_{ij}d_j = (\sum_{i=1}^{k} e_{ii})x(\sum_{i=1}^{k} e_{ii})$ for some k; thus, to prove the first assertion, we merely need to establish the following claim: if $e = \sum_{i=1}^{k} e_{ii}$, then $e\hat{R}e$ is the ring of central quotients of $B \cap e\hat{R}e$. Well there exist d_{ii} in D, such that $d_{ii}e_{ii} = e_{ii}d_{ii} \in B$, so $\Sigma e_{ii}d_{ii} = \Sigma d_{ii}e_{ii} \in B$. Let $x_1 = \Sigma e_{ii}d_{ii}$; then $ex_1 = x_1e = x_1$. Clearly x_1 is regular in $e\hat{R}e$, a simple artinian ring, so $x_1\hat{R}x_1$ is a left and right order in $e\hat{R}e$; thus, by the PI-theory, $e\hat{R}e$ is the ring of central quotients of $x_1\hat{R}x_1$. But $e\hat{R}e$ is the ring of central quotients of eRe, by proposition 7.6. Thus, $x_1\hat{R}x_1 = x_1e\hat{R}ex_1 \subseteq$ the ring of central quotients of $x_1eRex_1 = x_1Rx_1 \subseteq B \cap e\hat{R}e$, proving the claim. (ii). Suppose $r \in R \cap \text{soc}\hat{R}$ and $\overline{r_1r} \in \overline{\hat{R}r}$, and let $e = xr$ be an idempotent in $\hat{R}r$ whose image in $\overline{\hat{R}r}$ is 1. By (i), we can find nonzero μ in $R \cap Z$, such that $\mu er_1 \in R$ and $\mu x \in R$. Then $\overline{\mu e} \in \overline{Rr}$ and $(\overline{\mu e})(\overline{r_1r}) = (\overline{\mu er_1)r} \in \overline{Rr}$; it follows that $\overline{\hat{R}r}$ is the ring of central quotients of $\overline{Rr}$.

Theorem 7.8. Suppose R is a prime algie with proper GI and let B be an ideal of R, and suppose $f(X_1,\ldots,X_m) \in \hat{R}\{X\}$ such that (viewed in $\hat{R}$) $f(r_1,\ldots,r_m) = 0$ for all r_i in B. Then f is a GI of $\hat{R}$.

Proof. We shall only prove the assertion in the special case that f is completely homogeneous. (This situation covers all the applications we need here; the proof of the theorem then can be completed with a routine application of the Vandermonde argument of [25, theorem 3.1].) Well, let d_i be the degree of X_i in each monomial of f, for $1 \leq i \leq m$. By proposition 4.5, we are done if $f(x_1,\ldots,x_m) = 0$ for all x_i in soc$\hat{R}$. But, taking nonzero α_i in Z such that $\alpha_i x_i \in B$, we have $0 = f(\alpha_1 x_1,\ldots,\alpha_m x_m) = \alpha_1^{d_1} \ldots \alpha_m^{d_m} f(x_1,\ldots,x_m)$; thus $f(x_1,\ldots,x_m) = 0$.

Corollary 7.9. If R is prime with proper GI, then R and $\hat{R}$ are equivalent.

Let us conclude this section with an outline of what happens when R is a semiprime GPI-algie.

(1) R has left and right singular ideal 0. (Proof: If A is the left or right singular ideal of R, then, for any a in H(R)A, $\overline{Ra}$ is a semiprime PI-ring and thus has zero left and right singular ideals. But Ra $\subseteq$ H(R)A, implying a = 0. (This takes a bit of an argument along the lines of Fisher [10].) Thus H(R)A = 0, so A $\subseteq$ Ann(H(R)) = 0.

(2) It follows from (1) that R has a left maximal quotient ring $Q_1(R)$ and a right maximal quotient ring $Q_2(R)$, cf. Johnson [16].

(3) For any r in H(R), it is easy to check that $\overline{Q_1(R)r}$, being a semiprime PI-ring, is both the left and right maximal quotient ring of $\overline{Rr}$, and, being so, is Azumaya and von Neumann regular (cf. [10],

[16], and [26]). The left ideal $Q_1(R)r$ seems to take the place of minimal left ideals of $\hat{R}$. The most natural question in this context is, "Does $Q_1(R) = Q_2(R)$ always?" Another interesting question is, "What is the structure of a semiprime ring R with an element r such that $\overline{Rr}$ is a von Neumann regular, Azumaya PI-ring?" Note if such a ring R is prime, then R is strongly primitive. Indeed, $\overline{Rr}$ is a prime PI-ring which is von Neumann regular, so $Z(\overline{Rr})$ must be a field, implying $\overline{Rr}$ is central simple; thus, Rr is strongly primitive (by remark 7.4).

Appendix to §7.

Although Martindale's central closure is not needed in the general GI-theory of §§1-5, it is quite useful in tying GI-theory in with other structural assumptions. Smith [31] has given an example of such a procedure, and we shall give several other examples here. The proofs are not difficult, given the results in §7, and so they will be sketched.

Theorem 7.10. If R is a semiprime GPI-algie satisfying the ascending chain conditions on left annihilators, then R is a PI-ring, and the ring of central quotients of R is semisimple artinian.

Proof. It is standard to deduce that R is a finite subdirect product of prime algies with proper GI, each satisfying the ascending chain condition on left annihilators. Thus, it suffices to assume that R is prime. But, in view of the observations preceding proposition 7.6, $\mathrm{soc}\hat{R}$ cannot contain an infinite ascending chain of left ideals of $\hat{R}$ (since these are all annihilators), so $\hat{R}$ is simple artinian; it follows that $\hat{R}$ is a PI-ring, so R is a PI-ring.

Example 7.11. Let $P = \mathrm{End}_F M$ and let R be the subring of P generated by F and socP. Then socP is the only nontrivial ideal of R, so R satisfies the ascending chain condition on two-sided ideals. Also, R is primitive, with proper GI. However, R is not left or right Noetherian and is not a PI-ring. This situation cannot occur in PI-rings, because, by a recent result of Cauchon, every semiprime PI-ring satisfying the ascending chain condition on 2-sided ideals is left and right Noetherian.

A ring R has bounded index k if, for any nilpotent r in R, $r^k = 0$.

Theorem 7.12. If R is a prime algie with proper GI and with bounded index, then R is a PI-ring.

Proof. Let $\hat{R}$ be the central closure. By theorem 7.7, soc$\hat{R}$ has bounded index, implying soc$\hat{R}$ is central simple. Thus, $\hat{R}$ = soc$\hat{R}$ is a PI-ring, and therefore R is also a PI-ring.

In the proof of theorem 7.12 we could get a bound on the PI degree of R in terms of k and the height of f, as per the remark after theorem 5.5. Using [13, pp. 235-236], one can obtain

Corollary 7.13. Any algebraic, semiprime GPI-algie with bounded index is a PI-ring.

Corollary 7.13 leads us to Kurosh's problem - is every algebraic GPI-algie locally finite?

Martindale has discovered the following beautiful counterexample:

Let A be an F-algebra (without 1) which is nil but not locally nilpotent. (For example, we can use the famous Golod-Shafarecich algebra described in Herstein, Topics in Ring Theory, University of

Chicago press (1969).) Viewing A as left F-module, let $T = \mathrm{End}_F A$. T has a subring A_0 consisting of right multiplications by elements of A, and $A_0 \cong A$. Let R be the F-algebra of T generated by A_0 and socT. Clearly R is primitive, and every element of R is algebraic (since every element of R has a power in F+socT, which is algebraic). However, R is not locally finite because A is not locally finite.

Given $A \subseteq R$, let $C_R(A) = \{r \in R \mid [r,A] = 0\}$. We conclude with a theorem of Montgomery and Smith [22], viewed in terms of the above theory.

Theorem 7.14. If R is an algebra over a field K and if A is a finite dimensional central semisimple subalgebra such that $C_R(A)$ is a PI-ring, then R is a PI-ring. (We do not assume that the multiplicative unit of A is the multiplicative unit of R.)

Proof. If F is the algebraic closure of K then $A' = A \otimes_K F$ is "split" semisimple and $C_{R\otimes F}(A') = C_R(A) \otimes_K F$, a PI-ring; clearly we are done if we can show that $R \otimes_K F$ is a PI-ring. Thus we may assume that K is algebraically closed. Ib this case, we have two straightforward facts from [22]:

[22, lemma 2]. For every homomorphic image $\bar{R}$ of R, $C_{\bar{R}}(\bar{A}) = \overline{C_R(A)}$.

[22, lemma 3]. If R is prime and e is a primitive central idempotent of A, then $S_{2k}(eX_1e,\ldots,eX_{2k}e)$ is a GI of R, for suitable $k \leq$ PI degree of $C_R(A)$.

Suppose $C_R(A)$ has PI-degree k. If R is prime, then, by [22, lemma 3] and corollary 7.9, for any primitive central idempotent e of A, $S_{2k}(ex_1e,\ldots,eX_{2k}e)$ is a GI of $\hat{R}$, implying $\mathrm{rank}(e) \leq u$, where $u = \psi_0(2k)$ (cf. theorem 3.1). Now suppose 1 is a sum of t central idempotents. Then $\text{rank}(1) \leq tu$, implying $\text{rank}(A) \leq tu$. But $\text{rank}(C_R(A)) \geq \text{rank}(\text{soc}R) - \text{rank}(A)$. Hence, $\text{rank}(\text{soc}R) \leq tu + k$, so $\hat{R} = \text{soc}\hat{R}$ is a PI-ring, of degree bounded by a function of t and k.

If R is semiprime then we write R as a subdirect product of prime rings, each of which is PI of bounded degree, by [22, lemma 2] and the above paragraph. Hence R is a PI-ring.

Finally for arbitrary R, note that for any direct power ΠR, we can identify A with the "diagonal" $\{(a,a,\ldots) | a \in A\}$. The centralizer of A in ΠR is $\Pi C_R(A)$, a PI-ring; by [22, lemma 2] and the semiprime case, $(\Pi R)/\text{Nil}(\Pi R)$ is a PI-ring. Hence, by theorem 1.1, R is a PI-ring.

§8. Algies Without 1

Define a W-alg (or algie without 1) A to be a unital W-W bimodule with multiplication, satisfying every possible law of associativity and distributivity (of multiplication over addition) between elements of A and W. We shall describe a way of developing a GI-theory of algies without 1, which can be obtained by the previous results. As usual, "alg" means "W-alg".)

Lemma 8.1. Given an alg A, one can find an algie $A^{(1)}$ in which A is an ideal; if A is faithful as W-W bimodule, then $A^{(1)}$ is W-faithful.

Proof. The standard adjunction of 1 to a ring is modified slightly: Define $A^{(1)}$ to be the W-bimodule $W \oplus A$, with multiplication $(w_1,a_1)(w_2,a_2) = (w_1w_2, a_1w_2+w_1a_2+a_1a_2)$. Clearly $(1,0)$ is the multiplicative unit of $A^{(1)}$, and the map $W \to W\cdot(1,0)$ is a ring injection.

Identitfying A with $\{(0,a)|a \in A\}$, we see that A is an ideal in $A^{(1)}$. The last assertion is also clear.

There is a free W-alg, which we write as $(W\{X\})_0$; it is easy to see that $(W\{X\})_0 = \{f \in W\{X\}|f(0,\ldots,0) = 0\}$. An element f of $(W\{X\})_0$ is a GI of a W-alg A if f lies in the kernel of all W-alg homomorphisms from $(W\{X\})_0$ to A. This definition can be viewed in another way: If A is an ideal of an algie R, we say that an element f of $(W\{X\})_0$ is a GI of A if $f \in \cap\{\ker\phi | \phi: W\{X\} \to R$ is an algie homomorphism sending $(W\{X\})_0$ to $A\}$. The second definition makes sense for any alg, in view of lemma 8.1, and is easily seen to be equivalent to the first definition (and is thus well-defined). Using this definition, we can view a GI of an alg as a special case of a GI of an algie. One may be concerned that there are "enough" GI's for this viewpoint, but we already remarked in §2 that if f is a GI of R then $f-f(0,\ldots,0) \in (W\{X\})_0$ and is a GI of R, with the same nontrivial generalized monomials as f; hence, if R (as algie) has an R-proper (resp. strong R-proper) GI, then R (as alg) has an R proper (resp. strongly R-proper) GI. Equivalent (resp. m-equivalent) will now mean, satisfying the same generalized polynomials (resp. multilinear generalized polynomials) of $(W\{X\})_0$

At this point, it may be worth observing that any simple ring with 1, with nonzero socle, is simple artinian. Indeed, the socle is a nontrivial ideal and thus contains 1, and the assertion follows immediately. Since $\mathrm{soc}(\mathrm{End}_F M)$ is a simple non-artinian ring, we can expect the GI theory to be a little weaker for algies without 1. Indeed, theorem 5.2 and its consequences (notably theorem 6.1) fail to work unless we have 1. On the other hand, it is easy to notice that the constructions in §4 do not require the existence of 1, so we can lift proposition 4.5 and theorems 4.7 and 4.8 word for word, to obtain:

Proposition 8.2. If any alg R is a dense subring of End M_D, M a vector space over a division ring D, then R and End M_D are equivalent.

Theorem 8.3. Any prime alg R can be embedded in $End_F M$, for a suitable vector space M over a field F, which is m-equivalent to R.

Theorem 8.4. Any semiprime alg R can be embedded in $\Pi_{F_\gamma} End_{F_\gamma} M_\gamma$, for suitable vector spaces M_γ over fields F_γ, which is m-equivalent to R.

It is an easy matter to apply these theorems to §3 to obtain results analogous to those obtained in §6. One can refine the analysis a bit by modifying lemma 8.1.

Proposition 8.5. Given an alg R, let $R^{(1)}$ be as in lemma 8.1, and let B be an ideal of $R^{(1)}$, maximal with respect to $B \cap R = 0$. Define $R_1 = R^{(1)}/B$. R can be viewed as an ideal in R_1, and R intersects each nonzero ideal of R_1 nontrivially. If R is semiprime (resp. prime) then R_1 is semiprime (resp. prime).

Proof. Let $\overline{}$ denote the image in $R^{(1)}/B$. $\bar{R} = (R+B)/B \approx R/(R\cap B) = R$, so we identify R with $\bar{R}$. The next assertion is immediate. Now for any ideals $A_1 \supseteq B$, $A_2 \supseteq B$, of $R^{(1)}$ such that $\bar{A}_1\bar{A}_2 = 0$, we have $R \cap A_1A_2 = 0$, so $(R \cap A_1)(R \cap A_2) = 0$. If R is prime, or if $A_1 = A_2$ and R is semiprime, we have $R \cap A_1 = 0$, so $A_1 = B$. Thus, R semiprime (resp. prime) implies R_1 is semiprime (resp. prime).

Now if A is an ideal of a prime ring R, one can define the central closure $\hat{A}$ of A, as ring without 1, just as in §7, and it is easy to see that $\hat{A} = \hat{R}$. None of the proofs in §7 relied on the existence of 1, so we can apply corollary 7.9 to conclude that A is equivalent to $\hat{A} = \hat{R}$, which is equivalent to R. Thus, in conjunction with proposition 8.5, we have sketched a proof of

Theorem 8.6. For any prime alg R, R_1 (as constructed in proposition 8.5) is a prime alg equivalent to R.

References

[1] Amitsur, S.A., "Generalized polynomial identities and pivotal monomials," Trans. Amer. Math. Soc. 114 (1965), 210-226.

[2] ________, "Associative rings with identities," Some Aspects of Ring Theory, C.I.M.E. Conference, II Circlo August 23-31, 1965, pp. 1-48.

[3] ________, "Prime rings having polynomial identities with arbitrary coefficients," Proc. London Math. Soc. 17 (1967), 470-486.

[4] ________, "A note on PI-rings," Israel J. Math. 10 (1971), 210-211.

[5] ________, "On rings of quotients," Symposia Matematica vol. VIII, Instituto Nazionale di Alta Matematica, 1972.

[6] ________, and Levitzki, J., "Minimal identities for algebras," Proc. Amer. Math. Soc. 1 (1950), 449-463.

[7] Belluce, L.P., and Jain, S.K., "A remark on primitive rings and J-pivotal monomials," Journal of Mathematical Sciences 1 (1966), 49-51.

[8] ________, "Prime rings with a one-sided ideal satisfying a polynomial identity," Pacific Journal of Mathematics 24 (1968), 421-424.

[9] Erickson, T.S., Martindale, W.S., and Osborn, J.M., "Prime nonassociative algebras," Pacific J. Math. 60 (1975), 49-63.

[10] Fisher, J.W., "Structure of semiprime PI-rings," Proc. Amer. Math. Soc. 39 (1973), 465-467.

[11] Formanek, E., "Central polynomials for matrix rings," J. Algebra 23 (1972), 129-132.

[12] Kochen, S., "Ultraproducts in the theory of models," Annals of Mathematics 74 (1961), 221-261.

[13] Jacobson, N., "Structure of Rings," Amer. Math. Soc. Colloquium Publication XXXVII, (1964).

[14] ________, "PI-Algebras (An Introduction)," Lecture Notes in Mathematics 441 (1975), Springer-Verlag.

[15] Jain, S.K., "Prime rings having 1-sided ideal with PI coincide with special Johnson rings," J. Algebra 19 (1971), 125-130.

[16] Johnson, R.E., "Quotient rings of rings with zero singular ideal," Pacific J. Math. 11 (1961), 1385-1392.

[17] Kaplansky, I., "Rings with polynomial identity," Bull. Amer. Math. Soc. 54 (1948), 575-780.

[18] Kovacs, A., Dissertation (1975), Hebrew University of Jerusalem, Jerusalem, Israel.

[19] Martindale, W.S. III, "Prime rings satisfying a generalized polynomial identity," J. Algebra 12 (1969), 576-584.

[20] ________, "Primitive rings with involution whose symmetric elements satisfy a generalized polynomial identity," Proc. Amer. Math. Soc. 24 (1970), 508-511.

[21] ________, "Prime rings with involution and generalized polynomial identities," J. Algebra 22 (1972), 502-516.

[22] Montgomery, S., and Smith, M., "Algebras with separable subalgebras whose centralizer satisfies a PI," Communications in Algebra 3 (1975), 151-168.

[23] Posner, E., "Prime rings satisfying a polynomial identity," Proc. Amer. Math. Soc. 11 (1960), 180-184.

[24] Razmyslov, Ju. P., "On a problem of Kaplansky," Math. U.S.S.R. Izvestija 7 (1973) #3, 479-496.

[25] Rowen, L.H., "On rings with central polynomials," J. Algebra 31 (1974), 393-426.

[26] ________, "Maximal quotients of semiprime PI-algebras," Trans. Amer. Math. Soc. 196 (1974), 127-135.

[27] ________, "Generalized polynomial identities," J. Algebra 34 (1975), 458-480.

[28] ________, "Generalized polynomial identities II," J. Algebra 38 (1976), 380-392.

[29] ________, "Generalized polynomial identities III," (preprint).

[30] ________, "Polynomial identities of nonassociative rings," (preprint).

[31] Smith, M., "Rings with an integral element whose centralizer satifies a polynomial identity," (preprint).

A MECHANISM FOR DESCRIBING IDEALS IN GROUP RINGS

D. S. Passman*

University of Wisconsin at Madison
Madison, Wisconsin

Many of the interesting problems in ring theory are concerned with obtaining useful descriptions of certain ideals in a ring. For example, for any given ring, one certainly wishes to know what its Jacobson radical or prime radical looks like. In the case of group rings of infinite groups these problems have been studied with reasonable success over the past ten years or so. In this note we offer a mechanism for describing such ideals in group rings. This mechanism by no means solves any given problem, but rather it indicates a framework in which the answer should be described.

References for group ring material in general are the survey paper [2] and the books [1] and [3]. Proofs of most of the results mentioned below can be found in [3, Chapter 8].

1. Controlling Subgroups

Let K[G] denote the group algebra of G over the field K. If H is a subgroup of G, then there exists a natural projection map $\pi_H: K[G] \to K[H]$ given by

$$\pi_H\left(\sum_{x\in G} a_x x\right) = \sum_{x\in H} a_x x.$$

*Research supported in part by NSF Contract GP-32813X.

This is certainly not a ring homomorphism in general, but it is a right and left K[H]-module homomorphism.

If $H \triangleleft G$, then the conjugation action of G on H extends to an action of G on K[H]. If L is an ideal of K[H] we say that L is G-invariant if and only if $Lx = xL$ for all $x \in G$ and this holds if and only if $L = L^x = x^{-1}Lx$ for all $x \in G$. Certainly any characteristic ideal of K[H] is G-invariant.

The idea here is to try to describe ideals I of K[G] in terms of G-invariant ideals of K[H] with $H \triangleleft G$ and with H presumably smaller than G.

<u>Lemma 1.1</u>. Let I be an ideal of K[G] and let $H \triangleleft G$. Then

$$(I \cap K[H])\cdot K[G] \subseteq I \subseteq \pi_H(I)\cdot K[G].$$

Furthermore $I \cap K[H]$ and $\pi_H(I)$ are both G-invariant ideals of K[H].

We now see when equality can occur in the above.

<u>Lemma 1.2</u>. Let I be an ideal of K[G] and let $H \triangleleft G$. Then the following are equivalent.

(i) $I = (I \cap K[H])\cdot K[G]$.

(ii) $I = \pi_H(I)\cdot K[G]$.

(iii) $I = L\cdot K[G]$ for some G-invariant ideal L of K[H].

Furthermore when the above occurs we have

$$I \cap K[H] = \pi_H(I) = L.$$

When equality occurs, as in the previous lemma, we say that H controls I. This controlling concept is quite well behaved and indeed we have

<u>Lemma 1.3</u>. Let I be an ideal of K[G]. Then there exists a unique normal subgroup $\mathcal{C}(I)$, called the controller of I, with the following property. IF $H \triangleleft G$, then H controls I if and only if $H \supseteq \mathcal{C}(I)$.

In other words, in the lattice of normal subgroups of G, the controlling subgroups of I form an upwards cone with vertex $C(I)$. It seems reasonable that the first step in describing any ideal of K[G] should be the determination of its controller subgroup. Two special cases which are easy to compute are as follows. First, if I is the augmentation ideal $I = \omega(K[H])\cdot K[G]$, then as is to be expected we have $C(I) = H$. Second, if $I = eK[G]$ is generated by the central idempotent e, then $C(I) = \langle \mathrm{Supp}\ e \rangle$ is the finite normal subgroup of G generated by the support of e.

2. J-Controlling Subgroups

It is obvious that if I is a characteristic ideal of K[G], then $C(I)$ is a characteristic subgroup of G. Of course the determination of $C(I)$ is only the first step in the description of the ideal I. What is also needed is the determination of $I \cap K[C(I)]$. As it turns out, for the ideals we are most interested in, $I \cap K[C(I)]$ is usually equal to the Jacobson radical of $K[C(I)]$. This leads to the following definition.

Let I be an ideal of K[G] and let $H \triangleleft G$. Then we say that I is J-controlled by H if and only if

$$I = JK[H]\cdot K[G],$$

where JK[H] denotes the Jacobson radical of K[H]. Since JK[H] is a characteristic ideal of K[H], we conclude immediately that if the above occurs then H also controls I. Unfortunately, this J-controlling property is not as well behaved as ordinary controlling. For example it is not true that the intersection of two J-controlling subgroups of I also J-controls. Moreoever, it is not apriori true that if I has a J-controlling subgroup, then $C(I)$ also J-controls. This will be discussed further in the next section.

We remark that most of the considerations below are trivial in characteristic zero since characteristic zero group rings have no nonzero nil ideals. Thus in the following we will assume that char $K = p > 0$. In general the difficulty in obtaining $C(I)$ occurs in finding an upper bound for this subgroup. The following lemma is usually sufficient for the lower bound.

Lemma 2.1. Let I be an ideal of $K[G]$ and let W be a finite subgroup of G generated by p-elements. If $JK[W] \subseteq I$, then $W \subseteq C(I)$.

A simple consequence of this is that if G is finite then $C(JK[G]) = \mathbb{O}^{p'}(G)$, where the latter is the subgroup of G generated by all its p-elements. Furthermore, this subgroup J-controls $JK[G]$. Now we consider some results with G infinite.

If R is a ring, we let NR denote its nilpotent radical, so that NR is the join of all the nilpotent ideals of PR. Moreover we let PR denote the prime radical of R, that is PR is the intersection of all the prime ideals of R. In the group G we set

$$\Delta^p(G) = \langle x \mid x \text{ is a p-element with } [G:\mathbb{C}_G(x)] < \infty\rangle.$$

Then it can be shown that $\Delta^p(G)$ is a characteristic locally normal subgroup of G. The latter means that every finite subset of $\Delta^p(G)$ is contained in a finite normal subgroup of $\Delta^p(G)$ and in fact of G.

Theorem 2.2. Let char $K = p > 0$. Then $C(NK[G]) = \Delta^p(G)$ and the latter subgroup J-controls $NK[G]$.

Theorem 2.3. (Dyment and Zalesskii) Let char $K = p > 0$. Then there exists a certain locally finite characteristic p-subgroup P of G with $C(PK[G])/P = \Delta^p(G/P)$ and $C(PK[G])$ in fact J-controls $PK[G]$.

Furthermore, in either case, it is an easy matter to describe $JK[C(I)]$ in terms of the Jacobson radicals of the group rings of certain finite groups involved in $C(I)$.

3. The N*-Radical

What we really want is a description of the ideal JK[G]. However this problem has proved to be extremely difficult and computations have only been made for certain special classes of groups. In particular, we know the answer when G is solvable or when G is a linear group. In these cases it turns out that, when G is finitely generated, JK[G] = NK[G]. Motivated by this we introduce a new radical for rings.

Let R be a ring. We define

$$N^*R = \{\alpha \in R \mid \alpha S \text{ is nilpotent for all finitely generated subrings } S \text{ of } R\}.$$

It is easy to see that N*R is a characteristic nil ideal of R and furthermore if R is finitely generated as a ring then N*R = NR. For group rings we have

Lemma 3.1. Let K[G] be given. Suppose that JK[U] = NK[U] for all finitely generated subgroups U of G. Then JK[G] = N*K[G].

Based on this lemma and the above mentioned computations we conjecture that JK[G] = N*K[G] in general. Additional evidence for this conjecture is the following.

Theorem 3.2. Let K[G] be given. Then N*(K[G]/N*K[G]) = 0.

In other words, N* has radical-like behavior on group rings. Furthermore it can be shown by example that N* does not exhibit this behavior on rings in general. Now for the group G we define

$$\Lambda^p(G) = \langle x \mid x \text{ is a p-element with } [L:\mathbb{C}_L(x)] < \infty \text{ for all finitely generated subgroups } L \text{ of } G\rangle.$$

Then it can be shown that $\Lambda^p(G)$ is a characteristic locally finite subgroup of G and we have

Theorem 3.3. If char K = p > 0, then $\mathcal{C}(N^*K[G]) \subseteq \Lambda^p(G)$ and N*K[G] is J-controlled by $\Lambda^p(G)$ and by $\mathcal{C}(N^*K[G])$.

We remark that if the J = N* conjecture is correct, then some consequences of it and the previous result are as follows:

1. JK[G] is a nil ideal which is also equal to the Levitzki radical of K[G].
2. $\mathcal{C}$(JK[G]) is locally finite and J-controls JK[G].
3. If I is an ideal of K[G] which is J-controlled by some normal subgroup H, then $\mathcal{C}$(I) is a locally finite normal subgroup of G which J-controls I.
4. Certain questions as to when K[G] is semilocal would be solved in general.

We note that $\mathcal{C}(N^*K[G])$ has not been completely determined as yet. The problem here really amounts to a study of the Jacobson radical of group rings of locally finite groups and only partial results exist. Let G be a locally finite group. If A is a finite subgroup of G, we say that A is locally subnormal in G if and only if $A \lhd \lhd B$ for all finite subgroups B of G containing A. Then using these locally subnormal subgroups as building blocks, we can construct a certain characteristic subgroup $\mathcal{S}(G)$ of G. This subgroup has been studied in considerable detail and it has been shown to have a reasonably nice structure. One suspects that $\mathcal{S}(G)$ is the controller of JK[G] here and this is indeed true in all special cases which have been computed. The best result of any generality is

Theorem 3.4. Let G be a locally finite group and set $I = JK[\mathcal{S}(G)]\cdot K[g]$. Then $I \subseteq JK[G]$ and $N(K[G]/I) = 0$.

Finally, returning to arbitrary groups, if one puts together the two conjectures mentioned above, then one gets $C(JK[G]) = \mathcal{S}(\Lambda^p(G))$ and this group would J-control JK[G]. Furthermore the structure of the latter group is sufficiently simple to yield an adequate description of the Jacobson radical.

References

[1] A. V. Mihalev and A. E. Zalesskii, "Group rings," J. of Soviet Math. 4 (1975), 1-78.

[2] D. S. Passman, "Advances in group rings," Israel J. of Math. 19 (1974), 67-107.

[3] ________, "The Algebraic Structure of Group Rings," Interscience, New York, 1977.

LIE AND JORDAN MAPPINGS IN ASSOCIATIVE RINGS

Wallace S. Martindale, 3rd

University of Massachusetts
Amherst, Massachusetts

1. Introduction

Let R be an associative ring (not necessarily with 1). Under addition and the Jordan product xy + yx R is easily seen to be a Jordan ring; likewise under addition and the Lie product [x,y] = xy - yx R becomes a Lie ring. [R,R], the additive span of all commutators of R, and the center of Z of R are clearly Lie ideals of R. An important special case occurs when R is a ring with involution *, i.e. there exists an antiautomorphism * of period 1 or 2. In this case the set of symmetric elements $S = \{x \in R | x^* = x\}$ is a Jordan ring under xy + yx and $K = \{x \in R | x^* = -x\}$ is a Lie ring under xy - yx. Throughout this paper, without further mention, all rings are assumed to be 2-torsion free, i.e. 2x = 0 implies x = 0.

In the 1950's Herstein initiated a study of the Jordan and Lie structure of a simple ring R - a good account appears in [6]. The aim was to see how close R came to being simple as a Jordan ring and as a Lie ring, and, in case R had an involution, to study the Jordan and Lie simplicity of S and K. The inherently greater difficulties involved in the Lie situation were born out by the fact that, whereas S was shown to be a simple Jordan ring, the more complicated (but expected) result in the Lie case was that $[K,K]/[K,K] \cap Z$ was a simple

Lie ring (if (R:Z) > 16). Somewhat later, when extending results about simple rings to prime rings came into vogue, a systematic adaptation of Herstein's theorems to the prime ring case was made by Erickson [2].

At an AMS hour talk in 1961 [5] Herstein summarized what had been accomplished up to that point and listed a variety of problems to be tackled. Among these questions were several dealing with the characterization of additive mappings of R (and of S or K in case R had an involution) which preserved the Jordan or Lie product. It is our goal in the present article to give a brief account of some of the results on Jordan and Lie mappings that have been obtained during the past fifteen years. Also it is our intention to indicate how, in considering the Lie mapping problem, we were led naturally to the notions of extended center and central closure of a prime ring and to using the theory of generalized polynomial identities.

We were singularly unsuccessful when we first tackled these mapping problems for arbitrary simple rings without further assumptions. For instance, to this day Lie automorphisms of arbitrary division rings remain a total mystery to us. It wasn't until we lowered our signts and assumed that R contained some orthogonal idempotents that we were able to make any progress. Since all of our results take place in a framework of idempotents we close this introductory section with a brief account of this matter.

Suppose that R contains idempotents $e_1, e_2, \ldots, e_{n-1}$ whose sum is unequal to 1 (R may or may not contain 1) and such that $e_i e_j = 0$ for $i \neq j$. We formally set $e_n = 1 - (e_1 + \ldots + e_{n-1})$, let $R_{ij} = e_i R e_j$, and note that $R = \oplus \sum_{i,j=1}^{n} R_{ij}$. This is called the Pierce decomposition of R. In case R has an involution * and we take the e_i's to be symmetric idempotents it is immediate that $R_{ij}^* = R_{ji}$ for all i,j. Our point of view about the Pierce decomposition is that it is a device which enables one to "pretend" that R acts in certain ways as does the ring D_n of $n \times n$ matrices over a division ring D (where the e_i's may be taken as the n diagonal unit matrices).

2. Jordan Mappings

Let J be a Jordan subring of a ring R. A Jordan homomorphism ϕ of J into a ring R' is an additive mapping of J into R' such that $\phi(xy + yx) = \phi(x)\phi(y) + \phi(y)\phi(x)$ for all $x,y \in J$. It follows easily that $\phi(x^2) = \phi^2(x)$ and $\phi(xyz + zyx) = \phi(x)\phi(y)\phi(z) + \phi(z)\phi(y)\phi(x)$. Of particular interest are the cases when J = R or, when R has an involution, J is the set of symmetric elements S.

We begin by citing a theorem of Herstein [4]: if ϕ is a Jordan homomorphism of R onto a prime ring R' then ϕ is either a homomorphism or an antihomomorphism (the assumption of char. R $\neq$ 2,3 was later removed by Smiley). Since almost any ring is embeddable in a prime ring removal of the "onto condition holds little promise without other assumptions. Herstein's result has a special appeal for us in that, besides not requiring any chain conditions, it also does not require the presence of idempotents. We have been unsuccessful in obtaining any worthwhile result on Jordan mappings without assuming idempotents, although in passing we mention one instance, admittedly a very artificial one, where idempotents are not needed. Namely, let R and R' be simple rings with 1 with involutions of the first kind (i.e. leaving the centers elementwise fixed). Let ϕ be a "strong" Jordan homomorphism of S onto S' in the sense that ϕ also preserves "tetrads," i.e. $\phi(abcd + dcba) = \phi(a)\phi(b)\phi(c)\phi(d) + \phi(d)\phi(c)\phi(b)\phi(a)$ for all $a,b,c,d, \in S$. By Cohn [1] it follows that ϕ preserves all "n-trads"

$s_1s_2 \dots s_n + s_ns_{n-1} \dots s_1$. Then it can be shown (in a very few lines) that ϕ can be extended uniquely to an isomorphism of R onto R'.

Our main result on Jordan homomorphisms was obtained about a decade ago [15], and we state it as the following:

Theorem. Let R be a ring with involution containing three orthogonal symmetric idempotents e_1, e_2, e_3 whose sum is 1 and such that $Re_iR = R$, $i = 1,2,3$. If ϕ is a Jordan homomorphism of S into a ring R' then ϕ can be extended uniquely to a homomorphism of R into R'.

The condition $R = Re_iR$ is a very weak simplicity condition - it is satisfied by any matrix ring C_n, (and thus our theorem implies an earlier result of Jacobson and Rickart [11]) and also by von Neumann factors. The consequence of the theorem in Jordan algebra theory is that it says that the ring R itself is the special universal enveloping ring for the Jordan ring S. In low dimensions (in which the idempotent requirements are violated) there are counterexamples. For example, if R is the rational function field F(t) over the reals F, take $*: t \to -t$. Thus $S = F(t^2)$ and one may take $\phi: t^2 \to -t^2$. Let trivially, let $R = Q_2$, the 2×2 matrices over the quaternions and let * be conjugate transpose. Then S is a 6-dimensional Jordan algebra over the reals and, in terms of the obvious basis for S, we let ϕ be the identity mapping except for interchanging $j(e_{12}-e_{21})$ and $k(e_{12}-e_{21})$.

We touch very briefly on the proof of the theorem. It is immediate that the $\phi(e_i)$'s are idempotents in R' and one then shows they are orthogonal ones. This enables one to define an additive mapping from the "off-diagonal" Pierce space R_{ij}, $i \neq j$, into R' by $\sigma(x_{ij}) = \phi(e_i)\phi(x_{ij} + x_{ji})\phi(e_j)$, where $x_{ij} \in R_{ij} = e_iRe_j$, $x_{ji} = x_{ij}^* \in R_{ji}$, $i \neq j$. From $Re_iR = R$ one obtains $R_{ii} = R_{ij}R_{ji}$, $i \neq j$, so one may then attempt to extend σ to the "diagonal" Pierce spaces R_{ii}. The obvious well-definedness problem (stickiest part of the proof) succumbs to repeated application of the lemma that $\sigma(x_{ij})\sigma(x_{jk}) = \sigma(x_{ij}x_{jk})$, i.j.k distinct. By piecemeal computations on the various Pierce subspaces σ is finally shown to be a homomorphism which extends ϕ.

One useful corollary of the theorem is that it may be applied to the noninvolution situation where $\phi:R \to R'$ is a Jordan homomorphism. Indeed, suppose R contains three orthogonal idempotents e_1, e_2, e_3 whose sum is 1 and such that $Re_iR = R$. Then $A = R \oplus R^o$ (where R^o is the opposite ring of R) has the involution $*:(x,y) \to (y,x)$ symmetric elements $S = \{(x,x) | x \in R\}$, and orthogonal symmetric idempotents (e_i, e_i), $i = 1,2,3$ satisfying the conditions of the theorem. A Jordan homomorphism $\phi:R \to R'$ then induces a Jordan homomorphism $\psi:S \to R'$ according to $\psi(x,x) = \phi(x)$. By the theorem ψ can be extended to a homomorphism $\sigma:A \to R'$. Setting $\varepsilon_1:x \to (x,0)$ and $\varepsilon_2:x \to (0,x)$, one sees that $\phi = \sigma\varepsilon_1 + \sigma\varepsilon_2$ is a direct sum of a homomorphism and an anti-homomorphism of R into R'.

Another corollary is that a theorem of this type automatically implies an analogous result for Jordan derivations. Indeed, let R be a ring with involution (same conditions as in the theorem), and let λ be a Jordan derivation of S into R in the sense that λ is additive and $\lambda(st + ts) = \lambda(s)t + t\lambda(s) + s\lambda(t) + \lambda(t)s$. Then

$$\phi:s \to \begin{pmatrix} s & 0 \\ \lambda(s) & s \end{pmatrix}$$

is a Jordan homomorphism of S into R_2 and the theorem applies, yielding quickly the conclusion that λ can be extended to a derivation δ of R into itself.

We close our discussion of Jordan homomorphisms by raising several questions.

Can the condition on idempotents in our theorem be further weakened? One possibility would be to merely assume that the right annihilator of each Re_iR is 0. A more severe weakening would be to merely assume the existence of certain zero divisors, even in the case that R is simple.

Can the condition on idempotents be removed altogether, but with R assumed to be simple? For example, what if $R = R'$ is a division ring (dimension >4, of course) and ϕ is a Jordan automorphism of S? It would in fact make sense to first look at the easier question of Jordan derivations of S in this case.

The spectre of division rings makes us hesitate to suggest replacing the idempotent condition by a Noetherian condition. However, it might be of interest to investigate the situation where R, say, is an order (with involution) in D_n, D a division ring, $n \geq 3$.

3. Lie Mappings

Our interest in Lie mapping problems began with an aborted attempt at a dissertation topic; it was rekindled by Herstein's earlier mentioned hour talk [5], which drew our attention to the following result of Hua [8]. Let $R = D_n$, D a division ring, $n \geq 3$, and let ϕ be a Lie automorphism of R (in the obvious sense that ϕ is a one-one additive mapping of R onto itself such that $\phi(xy - yx) = \phi(x)\phi(y) - \phi(y)\phi(x)$ for all $x,y \in R$). Hua proved that $\phi = \sigma + \tau$, where σ is either an automorphism or the negative of an antiautomorphism of R and τ is an additive mapping of R into its center Z sending commutators to 0. As an illustration of such a Lie automorphism take $R = F_n$, F a field of characteristic 0, and $\phi = \sigma + \tau$, where σ is the identity map and $\tau(A) = (\text{trace } A)I$.

In [13] we were able to generalize Hua's theorem to the case where R was a primitive ring containing three nonzero orthogonal idempotents whose sum is 1. However, in the conclusion $\phi = \sigma + \tau$ we were forced to slightly enlarge the ring R so as to accommodate the images of σ and τ. An example (due to Kaplansky) can be used to illustrate this phenomenon. Let $F \subseteq E$ be fields such that E has an automorphism ρ which does not map F into itself, and let R be the set of all countably infinite matrices of the form $A + \lambda I$, where A is $n \times n$ over E, n varies, and $\lambda \in F$. Then it is easy to check that the Lie automorphism $\phi : A + \lambda I \to A^{\rho} + \lambda I$ cannot be written in the form $\sigma + \tau$, where the image of σ is contained in R. In retrospect this example was our first indication of the need for (what we later termed) the extended centroid and central closure of a ring; it wasn't until we generalized our Lie mapping theorem to prime rings [16] that we formalized those notions.

Let us pause momentarily to recall these notions for the reader. The extended centroid C of a prime ring R is simply the centroid of Utumi's maximal quotient ring Q [20] (see also [17]) and the central closure of R is the subring RC of Q generated by R and C. C is a field containing the centroid of R and RC is a prime algebra over C whose extended centroid is C. A prime algebra over a field Φ is closed if its extended centroid is Φ. For simple rings the centroid and extended centroid coincide, but for prime rings the centroid (even if it is already a field) may be properly contained in the extended centroid (as Kaplansky's example of the preceding paragraph illustrates). An important result for our purposes is that if A is a closed prime algebra over Φ then $A \otimes_{\Phi} A^{o} \simeq A_r A_{\ell}$, where A_r and A_{ℓ} are respectively the algebras of right and left multiplications determined by A.

We are now in a position to state our result on Lie isomorphisms of prime rings [16], [18]. Let T be a prime ring containing two nonzero orthogonal idempotents whose sum is 1, and let ϕ be a Lie isomorphism of T onto a prime ring R. Then ϕ is of the form $\sigma + \tau$, where σ is either an isomorphism or the negative of an anti-isomorphism of T into the central closure of A = RC of R and τ is an additive mapping of T into C sending commutators to 0.

The impact (admittedly a bit nebulous) of this theorem is that it reduces the classification problem for certain Lie algebras to associative theory. With this in mind, however, the motivation for our theorem is slightly off the mark, since the first of the four great classes of classical simple Lie algebras corresponds to [R,R] rather than to R, where $R = \Phi_n$, Φ the complexes. Hence the theorem of Howland [7] on Lie isomorphisms of [R,R] (with R suitably conditioned) is a better motivated result.

We turn now to the Lie isomorphism problem in rings with involution. Motivation is provided here in that the other three great classes of classical simple Lie algebras appear as the skew elements K (equal to [K,K] in this case) of Φ_n, Φ the complexes, under appropriate involutions. A result of Herstein (see [6]) says that if R is a simple ring with involution of the first kind then (modulo certain low dimensions) [K,K] is a simple Lie ring. It is still an open question whether K = [K,K] (for R simple with involution of the first kind). However, we are interested here in obtaining results for the situation when R is prime, and, as we shall presently see, [K,K] (and not K) is the appropriate Lie ring to consider.

If R is a prime ring with involution * then in a very natural way * can be extended to an involution of the central closure A = RC.

We define * to be of the first kind provided * induces the identity map on the extended centroid C. Kaplansky's example, cited earlier in this section, can easily be used to show that * may leave the center of R elementwise fixed but not be of the first kind. However, we restrict ourselves henceforth to involutions of the first kind. It is easy to find examples in the prime case where $K \neq [K,K]$ (e.g. take R to be the free algebra $\Phi[x,y]$ with $x^* = -x$ and $y^* = -y$). It follows from Erickson's paper [2] that [K,K] is itself a prime Lie ring, in the sense that for any two Lie ideals U,V of [K,K] $[U,V] = 0$ implies $U = 0$ or $V = 0$. We can therefore apply to [K,K] the results of some recent joint work with Erickson and Osborn [3], in which the extended centroid and central closure are defined for arbitrary nonassociative prime rings. Thus [K,K] has as its extended centroid Γ, C is clearly contained in Γ, and we wish to prove that $\Gamma = C$. We pursue this question further because it leads in a natural, if minor, way to the role played by generalized polynomial identities. Indeed, if $\lambda \in \Gamma$, pick $a \neq 0 \in [K,K]$ such that $b = \lambda a \subset [K,K]$, and note (1) $[[b,x],[a,x]] = 0$ for all $x \in [K,K]$. If (1) is "trivial" then a and b are C-dependent and so $\lambda \in C$; otherwise (1) is a nontrivial generalized polynomial identity satisfied by [K,K]. By our main result in [17] A must then be primitive with nonzero socle, which fact can then be used to force $C = \Gamma$.

We are now in a position to state our main result [19].

<u>Theorem</u>. Let R and T be prime rings with involutions of the first kind, with $A = RC$ and $B = TF$ respectively the central closures of R and T, K and L the skew elements of R and T, and $M = [K,K]$ and $N = [L,L]$. Assume furthermore that

(a) $(A:C) \neq 1,4,9,16,25,64$

(b) A contains two nonzero orthogonal symmetric idempotents e_1 and e_2 such that $e_1 + e_2 \neq 1$.

(c) For $i = 1,2$, $e_i \in \overline{M}_i$, the associative subring generated by $M_i = e_i A e_i \cap MC$

Then any Lie isomorphism ϕ of M onto N can be extended uniquely to an isomorphism of $\overline{M}$ onto $\overline{N}$.

Using Herstein's theorem that $\overline{[K,K]} = R$ when R is simple (see [6]) we have the following

<u>Corollary</u>. Let R and T be a simple rings with involutions of the first kind satisfying

(a) $(R:Z) \neq 1,4,9,16,25,64$, where Z is the centroid of R

(b) R contains two nonzero orthogonal symmetric idempotents e_1 and e_2 such that $e_1 + e_2 \neq 1$.

Then any Lie isomorphism of $[K,K]$ onto $[L,L]$ can be extended uniquely to an isomorphism of R onto T.

The need to avoid certain low dimensions, as indicated by condition (a), is illustrated by the following examples. For Φ the complexes let $R = \Phi_5$ with transpose as involution and let $T = \Phi_4$ with symplectic involution. Then the respective skew elements $M = K$ and $N = L$ are isomorphic 10-dimensional Lie algebras (see, e.g. [10], p. 142]. For $R = T = \Phi_8$ under transpose, there is a Lie automorphism of the skew elements M which cannot be extended to an automorphism of R (see, e.g. [10], Chapter 9, section 5). The finite dimensional case of the theorem has been well-known for a long time [10]. It was generalized by Klotz [12] to the situation where R and T were simple rings with 1 and R contained a certain finite dimensional simple subalgebra (with the same 1).

The proof of the theorem is rather lengthy, due in part to our desire to keep the argument self contained in the sense that no Lie theory is used. We touch upon only a few aspects of the proof. To begin with it is useful to be able to at least to show that the extended centroids C and F of R and T are isomorphic. Indeed, the extended centroids Γ and Λ of M and N are clearly isomorphic because ϕ is a Lie isomorphism. But the remarks made just prior to the statement of the theorem show that $C = \Gamma$ and $\Lambda = F$. The case where A or B (and hence A and B) satisfies a generalized polynomial identity involves considerable finite dimensional infighting - witness the 8×8 counterexample - but is technical in nature. The argument is shorter and more appealing in the case when neither A nor B satisfies a generalized polynomial identity. Here the key initial step is to prove that, if we set $N_i = \phi(M_i)$, $i = 1,2,3$, then $N_iN_j = 0$ for $i \neq j$. Indeed, pick $m_i \neq 0 \in M_i$, $i = 1,2,3$, set $a = \phi(m_1)$, $b = \phi(m_2)$, $c = \phi(m_3)$, and note that (2) $[[[n,a],b],c] = 0$ for all $n \in N$. Using [17] and the assumption that B does not satisfy a generalized polynomial identity we conclude that (2) must be "trivial" and so (3) $[[[y,a],b],c] = 0$ for all $y \in B$. Using the fact mentioned earlier in this section that $B \otimes_F B^o \simeq B_rB_\ell$ (3) may be translated into a tensor product equation, whence a short argument finally produces $ab = ac = bc = 0$.

If G is a Lie subring of a ring R then a Lie derivation $\mu: G \to R$ is an additive mapping μ of G into R such that $\mu([x,y]) = [\mu(x),y] + [x,\mu(y)]$ for all $x,y \in G$. Unfortunately the same device mentioned in section 1 for obtaining Jordan derivation theorems from Jordan homomorphism theorems cannot be used in the Lie situation, since the conditions on the image ring do not carry over. Some time ago we proved [14] that if R is primitive with an idempotent $e \neq 0,1$

then any Lie derivation $\mu: R \to R$ is of the form $\delta + \tau$, δ an ordinary derivation and τ an additive mapping into the center of R sending commutators to 0. Undoubtedly one should be able to replace primitivity by primeness.

More recently Jacobs, in his dissertation [9], established the following result. Let R be a simple ring with involution, having two nonzero orthogonal symmetric idempotents e_1, e_2 such that $e_1 + e_2 \neq 1$, and in addition, R not isomorphic to either the 4×4 matrices over a field or the 3×3 matrices over a field of characteristic 3. Then any Lie derivation μ of K into R is the restriction to K of $\delta + \tau$, where δ is a derivation of R into itself and τ is an additive mapping of R into its center sending commutators to 0. Counterexamples are given for the 3×3 and 4×4 matrix restrictions. Undoubtedly one should be able to show that simplicity in Jacob's theorem can be replaced by primeness.

By and large the open questions concerning Lie mappings are similar to those involving Jordan mappings.

Can our ever-present assumption of idempotents be removed? If one is looking for a counterexample one might look at the case that R and T are prime (perhaps integral domains) with involutions and ϕ is a Lie isomorphism of [K,K] onto [L,L]. On the other hand, if one is hoping for some kind of positive result, we would suggest looking at Lie derivations of a division ring. A positive result might also be obtained by studying Lie automorphisms of a free algebra - results of Cohn and Bergman might be applicable here.

Of lesser magnitude, but still of interest, would be to study the situation where R and T have involutions of the second kind (otherwise same conditions on R and T) and see if a Lie isomorphism of $[K,K]/[K,K] \cap C$ onto $[L,L]/[L,L] \cap F$ can be lifted to an isomorphism of $[\overline{K,K}]$ onto $[\overline{L,L}]$.

References

[1] P. M. Cohn, "On homomorphic images of special Jordan algebras," Canad. J. Math. 6 (1954), 253-264.

[2] T. S. Erickson, "The Lie structure in prime rings with involution," J. Algebra 21 (1972), 253-234.

[3] ________, W. S. Martindale, 3rd, and J. M. Osborn, "Prime non-associative algebras," Pac. J. Math. 60 (1975), 49-63.

[4] I. N. Herstein, "Jordan homomorphisms," T.A.M.S. 81 (1956), 331-341.

[5] ________, "Lie and Jordan structures in simple, associative rings," Bull. A.M.S. 67 (1961), 517-531.

[6] ________, "Topics in Ring Theory," University of Chicago Press, 1969.

[7] R. A. Howland, "Lie isomorphisms of derived rings of simple rings," T.A.M.S. 145 (1969), 383-396.

[8] L. Hua, "A theorem on matrices over an s-field and its applications," J. Chinese Math. Soc. (N.S.) 1 (1951), 110-163.

[9] D. R. Jacobs, "Lie derivations on the skew elements of simple rings with involution," Thesis, University of Massachusetts, 1973.

[10] N. Jacobson, "Lie algebras," Interscience tracts in pure and applied math., no. 10, Interscience, New York, 1962.

[11] ________ and C. E. Rickart, "Homomorphisms of Jordan rings of self-adjoint elements," T.A.M.S. 72 (1952), 310-322.

[12] E. A. Klotz, "Isomorphisms of simple Lie rings," T.A.M.S. 130 (1968), 337-352.

[13] W. S. Martindale, 3rd, "Lie isomorphisms of primitive rings," P.A.M.S. 14 (1963), 909-916.

[14] _______, "Lie derivations of primitive rings, Mich. Math. J. 11 (1964), 183-187.

[15] _______, "Jordan homomorphisms of the symmetric elements of a ring with involution," J. Algebra 5 (1967), 232-249.

[16] _______, "Lie isomorphisms of prime rings," T.A.M.S. 142 (1969), 437-455.

[17] _______, "Prime rings with involution and generalized polynomial identities," J. Algebra 22 (1972), 502-516.

[18] _______, "A note on Lie isomorphisms," Canad. Math. Bull. 17 (1974), 243-245.

[19] _______, "Lie isomorphisms of the skew elements of a prime ring with involution," Communications in Algebra, to appearl.

[20] Y. Utumi, "On quotient rings," Osaka J. Math. 8 (1956), 1-18.

THE GROUP RING OF THE INFINITE SYMMETRIC GROUP

E. Formanek

University of Chicago
Chicago, Illinois

Let F be a field of characteristic zero and let S_∞ denote the infinite symmetric group. Using the theory of Young diagrams for finite symmetric groups, the following theorems are proved:

Theorem 1. FS_∞ has precisely two maximal ideals.

Theorem 2. Any sum of prime ideals in FS_∞ is prime.

Theorem 3. Every ideal in FS_∞ is generated by a single element (as an ideal).

The talk is taken from joint work of the speaker and J. Lawrence. This appeared in the Israel Journal of Mathematics, 23 (1976), 325-331.

REPRESENTATIONS OF COMPLETE 2-DIMENSIONAL ORDERS

John H. Cozzens

Rider College
Trenton, New Jersey

Let R be a commutative Noetherian domain with quotient field K, D a finite-dimensional, central skewfield extension of K and $\Sigma = D_n$, a finite dimensional central simple K-algebra. A subring Λ of Σ is called an R-order in Σ if Λ_R is finitely generated and $\Lambda \otimes_R K \approx \Sigma$, i.e. Λ contains a K-basis for Σ_K. If Λ is an R-order in Σ and maximal with respect to this property, then Λ is called a maximal R-order in Σ. Clearly, R-orders and maximal R-orders are orders and maximal orders resp. in the more general sense (e.g. see Cozzens-Faith [75]).

When glb $\Lambda = 1$ and R is a DVR, the structure of Λ is well known. In particular, if R is complete, glb $\Lambda = 1$ iff Λ is conjugate to the set of all those matrices in d_n of the form

$$\begin{pmatrix} d & d & d & \cdots & d \\ m & d & d & \cdots & d \\ m & m & d & \cdots & d \\ \cdot & \cdot & \cdot & \cdots & \cdot \\ m & m & m & \cdots & d \end{pmatrix}$$

partioned into $n_i x n_j$ blocks all of whose entries belong to the indicated symbol (either d or m = rad d) where d is the unique maximal R-order in D (e.g. see Reiner [70]).

If glb $\Lambda = 2$ and R is a 2-dimensional regular local ring, the structure of Λ is reasonably well-known in at least 2 cases: when Λ is quasilocal; equivalently, maximal (Cozzens-Sandomierski [76], Fossum [68], Ramras [69], [74]), and when Λ is a maximal quaternion order (Riley [72]). Of particular interest to us is the following representation for p-connected (projectives are generators) Λ: if $\Sigma = D_n$, then $\Lambda \simeq d_n$ where d is a maximal R-order in D. Actually, the full force of p-connectivity is not used but rather the existence of a uniform progenerator.

Since an arbitrary R-free order is at least semilocal whenever R is local, there appears to be some hope of obtaining a reasonable representation (such as above) for an arbitrary 2-dimensional (global dimension) R-free, R-order Λ over a 2-dimensional regular local ring. Specifically one can ask: are such Λ p-connected whenever Λ is maximal?

The object of this paper is to summarize some joint work of Sandomierski and myself in the complete case and to discuss its implications to the aforementioned problem. Since the proof of the main theorem is very technical and involved, it is pointless to even attempt to sketch a proof. A detailed proof will appear elsewhere.

Henceforth, by an R-order, we shall mean a 2-dimensional, R-free R-order in a finite dimensional central simple K-algebra $\Sigma = D_n$ where K is the quotient field of R, a complete 2-dimensional regular local ring. Such an order is necessarily semiperfect by completeness of R. The terms order and maximal order will be reserved for the more general concepts.

Finally, I would like to express my thanks to the organizers of this conference for their generous hospitality.

The Main Theorem

Recall that an arbitrary R-order Λ in D_n is called tiled if Λ contains n orthogonal idempotents. By conjugating if necessary, we may assume that Λ contains the orthogonal idempotents e_{ii}, $i = 1,\ldots,n$, where $\{e_{ij} | 1 \leq i,j \leq n\}$ are the usual matrix units in D_n. We shall denote the set of all (i,j)-entries of Λ by Λ_{ij}. Λ is said to be reduced if $\forall_{i,j}$, $\Lambda_{ij}\Lambda_{ji} \subsetneq \Lambda_{ii}$. If $\Lambda_{ii} = d \,\forall i$, a fixed R-order in D, each Λ_{ij} is a fractional d-ideal of D which is invertible whenever d is maximal and Λ is R-free. We shall call Λ an integral frame if there exists an R-order $d \subset D$ such that $\Lambda \subseteq d_n$, and (upper) triangular if, in addition, $\Lambda_{ij} = d$ $\forall i \leq j$. Finally, if Λ is a triangular frame in d_n and $\Lambda_{ij} = p \;\forall i > j$, p a proper ideal of d, we shall denote this R-order by $\Delta_n(p)$.

Suppose Λ is an R-free, tiled R-order in D_n and each Λ_{ii} is a maximal R-order in D having the property that reflexive right ideals are projective. Then it is easy to show that Λ is conjugate to an integral frame in $(\Lambda_{11})_n$. This observation is used in the proof of the main theorem.

In a different direction, let Λ be any semilocal ring with maximal two-sided ideals M_i, $i = 1,\ldots,n$. A complete set of finite left localizations for Λ is a set $\{\Lambda \to \Gamma_i | i = 1,\ldots,n\}$, where each map $\Lambda \to \Gamma_i$ is a finite left localization of Λ at M_i i.e. each map $\Lambda \to \Gamma_i$ is a ring epi, $\Gamma_{i\Lambda}$ is finitely generated projective and $\Gamma_i \otimes_\Lambda \Lambda/M_j = 0$ $\forall\, j \neq i$. Λ is said to be (finitely) localizable if each Γ_i is also finitely generated projective as a left Λ-module.

Theorem 1. If Λ is non-maximal, then

1. D contains a 2-dimensional maximal (local) R-order and consequently, all maximal R-orders in D_n $\forall n \geq 1$ are quasilocal, 2-dimensional and conjugate.

2. Λ has a complete set of finite localizations (both left and right), each of which is maximal.
3. Λ is tiled.
4. Λ has no eventually idempotent maximal ideals iff Λ is Morita equivalent to $\Delta_m(p)$ for some invertible prime p of d, the (up to conjugation), unique maximal R-order in D.
5. If n is prime and Λ reduced, then Λ has no eventually idempotent maximal ideals. $\square$

Remark 1. When n is even and Λ is reduced, non-triangular examples do exist as the following example shows. However, the obvious generalization of this example to higher even n seems to be the only instance where Λ is not conjugate to a triangular frame.

Example 1. Let R be any 2-dimensional regular local ring with quotient field K,

$$\Gamma = \begin{pmatrix} R & R & R & R \\ q & R & q & R \\ p & p & R & R \\ pq & p & q & R \end{pmatrix} \quad p,q \text{ distinct}$$

invertible primes of R. Γ is an R-order in $\Sigma = K_4$ which is 2-dimensional $\Longleftrightarrow$ $p+q = m = \text{rad } R$. Otherwise $\text{glb } \Gamma = \infty$. Γ is not conjugate to any triangular frame by 1.4 since each maximal ideal of Γ is idempotent.

In a different vein, Γ is contained in precisely 4 maximal R-orders Γ_i, $i = 1,\ldots,4$, each of which is 2-dimensional, quasilocal and a finite left and right localization of Γ at some necessarily unique maximal ideal of Γ i.e. Γ is localizable. However, $\text{glb } \Gamma > \sup \text{glb } \Gamma_i$ whenever $p+q \subsetneq m$, providing a very simple example showing that Theorem 5.1 in Silver [67] does not extend to higher dimensions.

Remark 2. If Λ in Theorem 1 is in fact, maximal, then we do not know whether Λ is quasilocal. For, the proof of Theorem 1 relies heavily on the fact that the Λ in question is non-maximal. However, it seems likely that, in general, such a Λ is quasilocal.

Applications

Suppose R is no longer complete. If we set $\hat{R}$ = m-addic completion of R, m = rad R, $\hat{\Lambda} = \Lambda \otimes_R \hat{R}$ and $\hat{\Sigma} = \Sigma \otimes_K \hat{K}$, $\hat{K}$ = quotient field of $\hat{R}$, $\hat{\Lambda}$ is a 2-dimensional, $\hat{R}$-free $\hat{R}$-order in $\hat{\Sigma}$. Call an ideal I of Λ semilocal if Λ/I is semisimple. Clearly, Λ has no eventually idempotent semilocal ideals iff the same is true for $\hat{\Lambda}$. Now, if Λ is maximal and T is a non-trivial idempotent ideal of Λ, in particular, if T = trace P_Λ, P_Λ finitely generated projective, since $T^* = \Lambda$, T^* = Λ-dual of T_Λ, Ass (Λ/T) consists of maximal ideals (see Riley [65], 4.3 and Cozzens Sandomierski [76], 3.2). This readily implies that Λ/T is Artinian. Consequently, if I/T = rad (Λ/T). I is an eventually idempotent semilocal ideal of Λ. Hence, Λ has no non-trivial idempotent ideals iff Λ has no eventually idempotent semilocal ideals; in fact, as one can easily show, iff $\hat{\Lambda}$ has no eventually idempotent maximal ideals.

Now by Theorem 1, if $\hat{\Lambda}$ is non-maximal, $\hat{\Lambda}$ has no eventually idempotent maximals in many cases offering some evidence that the answer to our question in the introduction is "almost always" yes. However, if one can construct a 2-dimensional maximal R-order having example 1 or its analogue in higher dimensions as its completion, then such a Λ cannot be p-connected. Whether such an example already exists or, for that matter, can exist, is unknown to me at least.

Added in Proof The structure of Λ in Theorem 1 has now been explicitly determined in each of the remaining cases. Specifically, if n is even or if n is odd and composite, then Λ is either Morita equivalent to $\Delta_m(p)$ or to a ring of the form described in Example 1.

References

[60] Auslander, M. and Goldman, O., "Maximal orders," Trans. A.M.S. 97 (1960), 1-24.

[76] Cozzens, J. H., "Maximal orders and reflexive modules," Trans. A.M.S. 219 (1976), 323-336.

[75] ________ and Faith, C., "Simple Noetherian Rings," Cambridge University Press, 1975, Cambridge.

[76] ________ and Sandomierski, F. L., "Maximal orders and localization I," J. of Algebra, to appear.

[76] ________ and Sandomierski, F. L., "Reflexive primes, localization and primary decomposition in maximal orders," Proc. A.M.S., to appear.

[68] Fossum, R. M., "Maximal orders over Krull domains," J. of Algebra 10 (1968), 321-332.

[69] Ramras, M., "Maximal orders over regular local rings of dimension two," Trans. A.M.S. 142 (1969), 457-479.

[74] ________, "Orders with finite global dimension," Pacific J. Math. 50 (1974), 583-587.

[70] Reiner, I., "A survey of integral representation theory," Bull. A.M.S. 76 (1970), 159-227.

[65] Riley, J. A., "Reflexive ideals in maximal orders," J. of Algebra 2 (1965), 451-465.

[72] ________, "Maximal quaternion orders," J. of Algebra 23 (1972), 241-249.

[67] Silver, L., "Noncommutative localizations and applications, J. of Algebra 7 (1967), 44-76.

SOME RESULTS ON RINGS WITH FINITE GROUP ACTIONS

Joe W. Fisher
and
James Osterburg

University of Cincinnati
Cincinnati, Ohio

Introduction

Let R be a ring, G a finite group of automorphisms acting on R, and R^G the fixed subring of R. The objective of this paper is to use crossed products as a unifying approach to simultaneously obtain several results concerning finite group actions. More precisely let U denote the crossed product with trivial factor set of R with G. Then R can be made into a left U-module in such a way that the U-submodules of R are precisely the G-invariant left ideals of R. Moreover, R^G is isomorphic to $\mathrm{End}_U R$. Our main Theorem 1.3 asserts that if $|G|$ is invertible in R, then U and R form a relative projective pair, i.e. every short exact sequence of U-modules which splits over R also splits over U. As a corollary of this we obtain that R is a projective U-module. The stage is now set since $R^G \simeq \mathrm{End}_U R$ and R is a cyclic projective U-module. The play involves transferring properties of R to the endomorphism ring R^G and vice-versa. For example, the following theorem of Levitzki [10] opens the play: R is left Artinian implies R^G is left Artinian. Likewise, "left Artinian" can be replaced by "left Noetherian" or "left uniform dimension" or "left Krull dimension".

Another act, yields a theorem (Theorem 1.5) of Montgomery [13] which states that the Jacobson radical of R^G is contained in the Jacobson radical of R. This embodies examining the radical of R and then using well-known theorems concerning the radical of the endomorphism ring of a finitely generated projective module to calculate the radical of $\mathrm{End}_U R$ and hence R^G.

In the final act we reverse action to produce a result (Theorem 2.3) of Cohen-Montgomery [2]: If R^G is semisimple Artinian and R is semiprime, then R is semisimple Artinian. More can be obtained here, viz. if R^G is semiprimary, then R is semiprimary. Also we consider the case when R^G is perfect.

Our objective throughout this paper is to include as many related results, different proofs, enlightening examples, and questions as possible. Aside from the obvious disadvantages, it has the advantages of giving the reader an opportunity to see lots of techniques, examples, and problems.

Preliminaries-The Bergman-Isaacs' Theorem

Let R be an associative ring that contains a 1 (except in situations where R could not have a 1 such as R being nilpotent). Let G be a finite group of automorphisms of R and $R^G = \{r \in R \mid r^g = r \text{ for all } g \in G\}$. We assume R has no $|G|$ torsion, i.e. if $|G|r = 0$, then $r = 0$. The trace, tr(x), of an element x in R is $\Sigma_G x^g$. Clearly, tr(x) is an element of R^G for all $x \in R$. We call a left ideal L of R G-invariant if $L^g \subseteq L$ for each $g \in G$.

In 1935, J. Levitzki showed that if R is an Artinian ring, G a finite group of automorphisms such that R has no $|G|$-torsion and $R^G = 0$, then R is nilpotent [9, Theorem 4, p. 988]. In 1957, Higman showed if a ring R admits an automorphism of prime period leaving only zero fixed, then R is nilpotent [8]. The analogous result for finite groups was conjectured by Frobenius and shown by Thompson in [16]. Jacobson established the result for Lie rings in [11]. Recently Martindale and Montgomery have established an analogue for Jordan rings.

The associative ring case was extended by Herstein, Neumann [7], Sundstrom [15]. Bergman and Isaacs have the ultimate theorem. We will refer to it as the Bergman-Isaacs theorem [1, Proposition 2.3, p. 76].

Theorem. Let R be a ring, G a finite group of automorphisms of R such that R has no $|G|$-torsion. Then (1) if tr R is nilpotent then R is nilpotent, (2) if R is semiprime, then R^G is semiprime, (3) if L is a G-invariant left ideal of R, then either L is nilpotent or L contains nonzero elements of R^G.

Section 1. The Jacobson radical, the prime radical and the socle

By $J(\ )$ we mean the Jacobson radical of the ring in question. We are interested in the relationship between $J(R)$ and $J(R^G)$.

Lemma 1.1. $J(R) \cap R^G \subseteq J(R^G)$.

Proof. Let $x \in J(R) \cap R^G$, then $1 - sx$, $s \in R^G$, is invertible in R; hence in R^G.

We now distinguish three cases: R has $|G|$-torsion, i.e. there is $0 \neq r \in R$ such that $|G|r = 0$; R does not have $|G|$-torsion, i.e. if $|G|r = 0$, then $r = 0$; and even stronger $|G|$ is invertible in R.

Example 1. Let R be the 2×2 matrix ring over the field F of p elements, p a prime. Let g be the inner automorphism determined by $\begin{pmatrix} 1 & 1 \\ 0 & 1 \end{pmatrix}$. Then the cyclic group generated by g has order p. The fixed ring is $\{\begin{pmatrix} a & b \\ 0 & a \end{pmatrix}\ a,b \in F\}$. Thus $J(R^G) \not\subseteq J(R)$.

Thus in the modular case (i.e. R has $|G|$-torsion) $J(R) \cap R^G \neq J(R^G)$. On the other hand, S. Montgomery in [13] showed $J(R^G) \subseteq J(R)$, if $|G|^{-1}$ is in R. We will give another proof of this result, but first we ask

<u>Question 1.1</u>. Assume R has no $|G|$-torsion. Is $J(R^G) \subseteq J(R)$?

Let $G = \{1,h,\ldots,g\}$ and $|G|^{-1} \in R$. The <u>crossed product U of R with G</u> is the free left R-module generated by $u_1,u_h,\ldots,u_g$. By defining $ru_g \cdot tu_h = rt^g u_{gh}$, $r,t \in R$, $g,h \in G$, U is a ring. The unit element is u_1.

By sending r in R to ru_1 in U, we identify R with a subring of U. Thus any left U-module can be viewed as a left R-module. In this section, ${}_UM$ means M is a unital left U module. Similarly for ${}_RN$.

R is a left U module with the multiplication of $u = \Sigma t_g u_g \in U$ and $r \in R$ given by $\Sigma_G t_g r^g$. It is important not to confuse the ring multiplication of U and the module multiplication of R as a U-module.

In $\mathrm{End}_U R$, we will write our operators on the right. Let $t \in R^G$, by r_t we mean right multiplication by t. Since $t \in R^G$, r_t is a U homomorphism of R.

<u>Lemma 1.2</u>. The association t to r_t is a ring isomorphism from R^G to $\mathrm{End}_U R$.

<u>Proof</u>. This is routine. The 1 in R forces $t \to r_t$ to be onto.

Following D. Passman [14, p. 66] we say <u>U is R-projective</u> if N is a left U-submodule of a left U-module M such that N, when viewed as an R-module is an R direct summand of M, then N is a U direct summand. We assume $|G|^{-1} \in R$.

<u>Theorem 1.3</u>. U is R-projective.

Proof. Let ${}_U N$ be a submodule of ${}_U M$. The assumption that N is an R direct summand of M means there is an R map $h': M \to N$ such that $h'(n) = n$ for all $n \in N$.

Let $h(m) = |G|^{-1}\Sigma_G u_g h'(u_{g^{-1}} m)$. Then h is a U map from M to N such that $h(n) = n$ for all $n \in N$. Thus N is a U direct summand of M.

The reader interested in more details should consult [14, Lemma 15.2, p. 62].

Corollary 1.4. ${}_U R$ is a cyclic projective U-module.

Proof. Let $e = |G|^{-1}(u_1 + u_{g_1} + \ldots + u_{g_t})$, $e^2 = e \in U$. Now ${}_U R$ is isomorphic to ${}_U Re$, a left U-submodule of U that is an R direct summand of U. Hence a U direct summand. Ergo ${}_U R$ is projective. Since ${}_U R$ is generated by 1, R is cyclic.

We now consider various radicals. By $J({}_U R)$ we mean the intersection of the maximal left U-submodules of R, i.e. the intersection of the maximal G-invariant left ideals of R. As before, $J(R^G)$, $J(R)$ and $J(U)$ mean the Jacobson radical of these rings.

Now the corollary and [5, Proposition 17.10, p. 196] together imply $J({}_U R) = J(U)\cdot R$. Furthermore, since U is R projective $J(U) = J(R)\cdot U = U\cdot J(R)$ [14, Theorem 16.3, p. 65]. Combining these results we have $J({}_U R) = (J(R)\cdot U)R = J(R)(U\cdot R) = J(R)\cdot R = J(R)$ (Recall the module multiplication!).

What is the Jacobson radical of $End({}_U R)$? Since ${}_U R$ is f.g. and projective, $J(End({}_U R)) = Hom({}_U R, J({}_U R))$ by [5, Corollary 17.12, p. 197]. Hence, using the preceding paragraph, we have $J(End({}_U R)) = Hom({}_U R, J(R))$. We now give another proof of [13, Theorem 7].

Theorem 1.5 (Montgomery [13]). $J(R^G) \subseteq J(R)$. Hence, $J(R) \cap R^G = J(R^G)$.

Proof. Let $t \in J(R^G)$ by Lemma 1.2, $r_t \in J(\mathrm{End}(_U R))$. So $r_t\colon {}_U R \to J(R)$ but $r_t(1) = t \in J(R)$. Ergo $J(R^G) \subseteq J(R)$.

Let B be a G-invariant left ideal of R. Each $g \in G$ induces an automorphism $\overline{g}$ of $\overline{R} = R/B$, as follows, $g(r + B) = g(r) + B$. The association g to $\overline{g}$ is a group homomorphism from G to $\mathrm{Aut}(\overline{R})$. The kernel H is a normal subgroup of G. Let $\overline{G} = G/H$.

Lemma 1.6. Assume there is $d \in R$ such that $\mathrm{tr}\, d - 1 \in J(R)$. Then $(R/B)^{\overline{G}} = R^G + B/B = R^G/B \cap R^G$.

Proof. Let $d \in R$ and $\mathrm{tr}\, d - 1 \in J(R) \cap R^G \subseteq J(R^G)$, i.e. $\mathrm{tr}\, R + J(R^G) = R^G$, but $J(R^G)$ is small in R^G. So there is $c \in R$ such that $\mathrm{tr}\, c = 1$. Now let $x + B \in (R/B)^G$, then $x + B = x\cdot 1 + B = x\, \mathrm{tr}\, c + B = \mathrm{tr}\,(cx) + B \in R^G + B/B$. Thus $\overline{R}^{\overline{G}} = \overline{R^G}$.

To use Bergman-Isaacs' theorem on factor rings we must rule out $|G|$-torsion. If we assume R modulo every G-invariant ideal has no $|G|$-torsion then $|G|$ is invertible in R. For if not, $|G|R$ is a proper G-invariant ideal of R, but $R/|G|R$ certainly has $|G|$-torsion. Furthermore, $|G|^{-1}$ is an element of trace 1. In conclusion, if $|G|^{-1} \in R$, then for a G-invariant ideal B we have (i) $\overline{R} = R/B$ has no $|G|$-torsion and (ii) $\overline{R}^{\overline{G}} = \overline{R^G}$.

Theorem 1.7. Let $P(R)$ and $P(R^G)$ be the prime radicals of R and R^G. If $|G|^{-1} \in R$, then $P(R) \cap R^G = P(R^G)$.

Proof. If P is a prime ideal in R and $g \in G$, then P^g is a prime ideal also. Consequently, $P(R)$ is G-invariant since it is the intersection of all the prime ideals of R. If $\overline{R} = R/P(R)$, then by Lemma 1.4, $\overline{R}^{\overline{G}} = R^G/R^G \cap P(R)$. By Bergman-Isaacs' theorem, $\overline{R}^{\overline{G}}$ is semiprime, hence $R^G \cap P(R) \supseteq P(R^G)$.

In order to show $R^G \cap P(R) \subseteq P(R^G)$, we use an elementwise characterization of the prime radical given in Lambek [9, Proposition 1, p. 56], i.e. $P(R) = \{a \in R | a = a_0, a_1, a_2, \ldots$ with $a_{n+1} \in a_n R a_n$ is ultimately zero$\}$. Let $a \in P(R) \cap R^G$, $a = a_0, a_1, a_2$, be a sequence of elements from R^G, where $a_{n+1} \in a_n R^G a_n$. Since $a_n \in P(R)$, this sequence is ultimately zero. Thus $a \in P(R^G)$.

We begin our discussion of the socle with

Example 1.2. Let F be a field, char $F \neq 2$, $R = \{\begin{pmatrix} a & b \\ 0 & c \end{pmatrix} | a,b,c \in F\}$. Take $g\begin{pmatrix} a & b \\ 0 & c \end{pmatrix} = \begin{pmatrix} a & -b \\ 0 & c \end{pmatrix}$, $g^2 = 1$. $R^{<g>} = \{\begin{pmatrix} a & 0 \\ 0 & c \end{pmatrix} | a,c \in F\}$. Now R is semiprimary, so the left socle is the right annihilator of J, i.e. $\{\begin{pmatrix} 0 & b \\ 0 & 0 \end{pmatrix} | b \in F\}$. The right socle is the left annihilator of J, i.e. $\{\begin{pmatrix} 0 & b \\ 0 & c \end{pmatrix} | b,c \in R\}$. Neither l. soc$(R) \cap R^G$ nor r. soc$(R) \cap R^G$ contains soc(R^G).

In the next two lemmas we just assume R has no G-torsion.

Lemma 1.8. Let R be semiprime and E an essential left ideal of R. Then E contains a nonzero element of R^G. In fact, $\cap_G E^g \cap R^G$ is an essential left ideal of R^G.

Proof. Let E be an essential left ideal of R. Then E^g is essential for all $g \in G$. Consequently, $\cap_G E^g$ is an essential G-invariant left ideal of R. By Bergman-Isaacs' theorem $\cap_G E^g \cap R^G$ is nonzero. Moreover, it is essential as a left ideal of R^G. Let $0 \neq B$ be a left ideal of R^G. So RB is a nonzero G-invariant left ideal of R, so $RB \cap (\cap_G E^g) \neq 0$. Now $\mathrm{tr}(RB \cap (\cap_G E^g)) \neq 0$, otherwise $RB \cap (\cap_G E^g)$ would be nilpotent (Bergman-Isaacs, again), hence 0. But
$0 \neq \mathrm{tr}(RB \cap (\cap_G E^g)) \subseteq RB \cap (\cap_G E^g) \cap R^G \subseteq B \cap (\cap_G E^g \cap R^G)$.

Lemma 1.9. Let R be semiprime. Then soc $R \cap R^G \supseteq$ soc R^G.

We offer two proofs.

1) By a result of Kasch-Sandomierski [5, Proposition 9.7, p. 118] soc(R) is the intersection over all essential left ideals E of R. So $\mathrm{soc}(R) = \cap_E(\cap_G E^g)$ and $\mathrm{soc}\, R \cap R^G$ is an intersection of essential left ideals of R^G. Hence $\mathrm{soc}\, R \cap R^G \supseteq \mathrm{soc}\,(R^G)$.

2) We know R^G is semiprime by Bergman-Isaacs' theorem. Let $R^G e$, $e^2 = e$, be a minimal left ideal of R^G. If Re contains a left ideal L such that $\cap_G L^g \neq 0$, then $0 \neq \cap_G L^g \cap R^G \subseteq Re \cap R^G = R^G e$. So $e \in \cap L^g$, $e \in L$ and $Re = L$. Thus we assume $\cap_G L^g = 0$ for all left ideals L of R contained in Re.

By Zorn's lemma, among the left ideals of R in Re there is a maximal one L, i.e. Re/L is a simple R-module. By the above, we have $\cap_G L^g = 0$. So we can embed Re in $\Sigma\oplus_G Re/L^g$, a finite direct sum of simples. Thus Re is an Artinian left R module. Let $0 \neq M$ be a minimal left ideal of Re. Now $\Sigma_G M^g = Re$, since Re cannot contain a proper nonzero G-invariant ideal. Thus $e \in \mathrm{Soc}(R)$, so $\mathrm{soc}(R^G) \subseteq \mathrm{soc}\, R \cap R^G$.

By Z(R) we mean the singular ideal of R. Lemma 1.7 is "essentially" the proof of $Z(R) \cap R^G \subseteq Z(R^G)$. Thus an application of the Bergman-Isaacs' theorem shows, if R is semiprime with no $|G|$-torsion and $Z(R^G) = 0$, then $Z(R) = 0$.

<u>Question 1.2</u>. Let R be semiprime. Is $\mathrm{soc}\, R \cap R^G$ contained in $\mathrm{soc}\,(R^G)$? Is $Z(R^G)$ contained in $Z(R) \cap R^G$?

2. <u>Perfect Rings</u>

For many of the following theorems we will give two proofs. When we do this, one proof will involve the crossed product. This provides a unified approach. We begin with an old theorem of J. Levitzki [10, Theorem 4, p. 988].

Theorem 2.1. Let R be a left Artinian ring, G a finite group of automorphisms of R such that R has no $|G|$-torsion. Then R^G is left Artinian.

Proof. The order of G is a regular element of R; hence, invertible in R. So by Corollary 1.4, $R^G = \text{End}_U R = eUe$, $e^2 = e$ an element of U. Now U being a finitely generated R module, is left Artinian. Hence, R^G is left Artinian.

We give an example to illustrate the importance of the lack of torsion assumption in Theorem 2.1.

Example 2.1. Let $K = Z_2(\{t_i\})$ the rational functions in a countable number of indeterminants over the field of 2 elements. Let $R = K + Kx + Ky$, $0 = x^2 = y^2 = xy = yx$. Now R is Artinian and $J(R) = Kx + Ky$.

The field K has derivation D, $D(f) = \Sigma t_{i+1} \partial f/\partial t_i$, let $K_0 = \{x \in K | D(x) = 0\}$, then $[K:K_0]$ is infinite. We let $g(a + bx) = a + (Da + b)x$ $g(y) = y$, so $g^2 = 1$. If we let $G = \langle g \rangle$, then R has 2 torsion and $R^G = K_0 + Kx + Ky$, which is not Artinian.

Now let $h(x) = x + y$, $h(y) = x$ and $h(a) = a$, $a \in K$. Then $h^3 = 1$, $G = \langle h \rangle$ has order 3. So R has no $|G|$-torsion and $R^G = K$.

The left Krull dimension of R, denoted by $K \dim R$, is defined by transfinite recursion as follows, if $R = 0$, then $K \dim R = -1$, if α is an ordinal and $K \dim R < \alpha$, then $K \dim R = \alpha$ provided that there is no infinite descending chain $R = L_0 \supseteq L_1 \ldots$ of left ideals L_i such that $K \dim(L_{i-1}/L_i) < \alpha$. For a definitive account of Krull dimension, see Gordon-Robson [6].

Theorem 2.2. Let R be a ring with Krull dimension α and $|G|^{-1} \varepsilon R$, then R^G has Krull dimension less than or equal to α.

Proof. U being a finitely generated R module has Krull dimension less than or equal to $K\ \dim_R R$. By [6, Proposition 1.2, p. 6] and $_UR$ finitely generated and projective [Corollary 1.4], we have
$K\ \dim\ \mathrm{End}_U R \leq K\ \dim_U U \leq K\ \dim_R U \leq K\ \dim_R R$. By Lemma 1.2,
$K\ \dim\ R^G \leq K\ \dim_R R$.

Now some examples to show in the presence of $|G|$-torsion, Theorem 2.2 fails.

Example 2.2. Let K be a field of characteristic 2. Define $R = Ke + Ka$, e is the unit of R, $a^2 = 0$ and $ak = \phi(k)a$, ϕ an endomorphism of K. Put $K_0 = \{k \varepsilon K | \phi(k) = k\}$. Now $e + a$ determines an inner automorphism of R of period 2. Clearly, $R^G = K_0 e + Ka$. Now $K\ \dim R = 0$, but if $[K:K_0] = \infty$, R^G does not have finite uniform dimension; hence, does not have Krull dimension [6, Proposition 1.4, p. 7].

Another example is, let Z_2 be the field of 2 elements and R be $\begin{pmatrix} Z_2(x) & Z_2(x) \\ 0 & Z_2 \end{pmatrix}$. For g take conjugation by $\binom{1\ 1}{0\ 1}$. Then $g^2 = 1$ and $R^G = \{\binom{a\ b}{0\ a} | a \varepsilon Z_2,\ b \varepsilon Z_2[x]\}$. Now R has left Krull dimension by [6, Proposition 1.3, p. 7], but R^G does not have finite uniform dimension; hence, R^G does not have Krull dimension.

Question 2.1. Let R have Krull dimension α, G a finite group of ring automorphisms such that R has no $|G|$-torsion. Does R^G have Krull dimension $\leq \alpha$?

Jondrup in [12] has noted that if $|G|^{-1} \varepsilon R$ and R is left self-injective, then R^G is left self-injective. Combined with Levitizki's theorem we have if R is quasi-Frobenius and R has no

$|G|$-torsion, then R^G is quasi-Frobenius. This leads us to ask

Question 2.2. Pick your favorite type of Artinian ring. Is R^G favored also? Perhaps, quasi-favored?? What happens if R has $|G|$-torsion?

We now turn the tables and consider conditions that "go up". We begin with a result of Cohen and Montgomery [2, Theorem, p. 189]. For this theorem, we do not need to assume R has an identity.

Theorem 2.3. Let R be a semiprime ring and G a finite group of automorphisms of R such that R has no $|G|$-torsion. Then if R^G is semiprime Artinian, then R is semiprime Artinian.

Proof. We begin by showing as Cohen-Montgomery did, that the identity of R^G, say e, is the identity of R. Let $L = \{re-r \mid r \in R\}$, L is G-invariant and $L \cap R^G = 0$, by Bergman-Isaacs' theorem, L is nilpotent. But R is semiprime which forces e to be the identity of R. Also, $|G|$ is invertible in R, as R^G satisfies the DDC. We will give three proofs of this result. The first involves the crossed product, U.

We begin by noting U is semiprime. For $J(U) = J(R)U = 0$ by Theorem 1.5 and [14, Theorem 16.3, p. 65]. We know $\mathrm{End}_U R$ is semi-simple Artinian, since $R^G \simeq \mathrm{End}_U R$ by Lemma 1.2. By Corollary 1.4, $_U R$ is finitely generated projective. It is not hard to show that $_U R$ is completely reducible. For example, see [17, Proposition 5.2, p. 254].

We now argue that R is completely reducible as a left R module. Equivalently, R has no proper essential left ideals [9, Proposition 2, p. 61]. To this end, let E be an essential left ideal of R. Then $\cap_G E^g$ is an essential G-invariant left ideal of R. But R has no proper essential U submodules. Hence, $\cap_G E^g$ is either R or zero. In either case, we have a contradiction. Thus, $_R R$ is completely reducible.

The second proof uses Lemma 1.9. By hypothesis $R^G = \text{soc } R^G$, but $\text{soc}(R^G) \cap \text{soc } R$. As R^G contains 1, we must have soc R = R. So we have the result.

Now for our last proof. Let E be an essential left ideal of R. By Lemma 1.8, $\cap_G E^g \cap R^G$ is an essential left ideal of R^G. So $_G E^g \cap R^G$ is either R^G or zero. In the first instance, E = R, in the second instance, E = 0. Thus, R contains no proper G-invariant ideals. Since J(R) = 0, we have the result.

A reasonable expectation would be R^G Artinian, then R Artinian. The next examples show this is false. Of course, neither is semiprime.

<u>Example 2.3</u>. Let R be the ring of all 2 × 2 matrices $\begin{pmatrix} a & b \\ 0 & c \end{pmatrix}$, a rational and b, c real. Then R is not left Artinian, but semiprimary, (i.e. R/J(R) Artinian, and J(R) nilpotent). Let $g\begin{pmatrix} a & b \\ 0 & c \end{pmatrix} = \begin{pmatrix} a & -b \\ 0 & c \end{pmatrix}$, $g^2 = 1$ and $R^{<g>} = \begin{pmatrix} a & 0 \\ 0 & c \end{pmatrix}$ a semiprimitive Artinian ring.

Let F be a field char F ≠ 2 and V an infinite dimensional vector space over F. Let V be a ring with zero multiplication and R be all pairs (v,f), v ε V, f ε F. Let (v,f)(v',f') = (f'v+fv',ff') and g(v,f) = (-v,f). Then R^G = F, but R is not Artinian. This example is commutative.

<u>Theorem 2.4</u>. Let R be a ring, G a finite group of automorphisms of R such that R has no |G|-torsion and R^G semiprimary. Then R is semiprimary.

<u>Proof</u>. Since R^G is semiprimary, it satisfies the DCC on principal left ideals. Thus |G| is invertible in R. Now $J(R) \cap R^G = J(R^G)$ by Theorem 1.5. We give a short proof of this using Bergman-Isaacs' theorem. Since $J(R) \cap R^G \subseteq J(R^G)$, J(R) is nilpotent, by Bergman-Isaacs'

theorem. Now G acts on $RJ(R^G)$ and $tr(RJ(R^G)) \subseteq J(R^G)$. Again by Bergman-Isaacs', $RJ(R^G)$ is nilpotent. Thus $J(R^G) \subseteq RJ(R^G) \subseteq J(R)$. So $J(R) \cap R = J(R^G)$. By Theorem 2.3, $R/J(R)$ is Artinian. Thus R is semiprimary.

As always, R has no $|G|$-torsion.

Theorem 2.5. a) If R^G is semiprimary such that $R^G/J(R^G)$ is simple then R is semiprimary and $R/J(R)$ is a finite direct sum of $\leq|G|$ isomorphic simple rings.

b) If R^G is semiprimary and $R^G/J(R^G)$ is a division ring, then R is semiprimary and R/J is a direct sum of $\leq|G|$ isomorphic division rings.

Proof of a). Let B be a maximal 2-sided ideal of R/J. Now $\cap_G B^g$ has zero intersection with $\overline{R}^G = R^G/J(R^G)$. By Bergman-Isaacs', $\cap B^g$ is nilpotent, hence 0. Thus R can be embedded in $\Pi_G R/B^g$. Proof of b). Just like a), but let B be a maximal left ideal.

Theorem 2.6. Let R be a ring, G a finite group of automorphisms of R such that R has no $|G|$-torsion. Then if R has DCC on G-invariant left ideals, R is semiprimary. Moreover, if R is semiprime, then R is left Artinian.

Proof. For each k, the ideal $R|G|^k$ is G-invariant. Thus the chain $R|G| \supseteq R|G|^2 \ldots$ eventually terminates. So $|G|$ is invertible in R. Thus for any left ideal L of R^G, $RL \cap R^G = L$. So R^G is left Artinian and by Theorem 2.4, R is semiprimary.

Unfortunately, we do not have the same kind of results for left perfect rings. The difficulty is that T-nilpotence does not go up. So we ask if there is a T-nilpotent version of the Bergman-Isaacs' theorem.

Of course, neither R nor R^G can have a 1.

Question 2.3. Let R^G be left T-nilpotent. Is R left T-nilpotent?

A positive answer to this question would yield: if R^G is left perfect, then R is left perfect. We can prove this if R^G is contained in the center of R. We give some examples of this behavior. See also the last section of [1] and [4] for more about this assumption.

Example 2.4. Let R be a commutative ring, G a finite group of automorphisms of R such that R has no $4|G|$-torsion. Let A be the 2×2 matrices over R, then $g\begin{pmatrix} a & b \\ c & d \end{pmatrix} = \begin{pmatrix} a^g & b^g \\ c^g & d^g \end{pmatrix}$, $g \in G$, defintes an automorphism of A. So A^g is the 2×2 matrices over R^G. Now let Z_2 be the cyclic group of order 2 and $H = Z_2 \times Z_2$. On R^G define $(1,0)\begin{pmatrix} a & b \\ c & d \end{pmatrix} = \begin{pmatrix} a & -b \\ -c & d \end{pmatrix}$, $(0,1)\begin{pmatrix} a & b \\ c & d \end{pmatrix} = \begin{pmatrix} d & c \\ b & a \end{pmatrix}$ and $(1,1)\begin{pmatrix} a & b \\ c & d \end{pmatrix} = \begin{pmatrix} d & -c \\ -b & a \end{pmatrix}$. Thus the fixed ring of $G \times H$ is $\begin{pmatrix} a & 0 \\ 0 & a \end{pmatrix}$, $a \in R^G$. Thus the fixed ring is properly contained in the center.

A similar example would be the integral quaternions and let G be the inner automorphisms determined by i,j and k. Then the fixed ring coincides with the center.

Finally, let R be a commutative perfect ring, M a R-module. Take for A all matrices of the form $\begin{pmatrix} r & m \\ 0 & r \end{pmatrix}$, $r \in R$, $m \in M$. $g\begin{pmatrix} r & m \\ 0 & r \end{pmatrix} = \begin{pmatrix} r & -m \\ 0 & r \end{pmatrix}$. Then $R^G = R$.

Theorem 2.7. Assume the finite group G acts on R without $|G|$-torsion. If R^G, a perfect ring, is contained in the center of R, then R is left and right perfect. Thus if R is commutative with perfect fixed ring, then R is perfect.

Proof. By Theorem 1.5, $RJ(R^G) \subset J(R)$ and our central hypothesis makes $RJ(R^G)$ a 2-sided G-invariant ideal of R. Now G acts on $R/RJ(R^G)$

with fixed ring $R^G/J(R^G)$. By Theorem 2.4, $R/RJ(R^G)$ is semiprimary. Thus $J(R)^k \subseteq RJ(R^G)$, for some k.

One characterization of left T-nilpotence is that for any R^G-module, hence for any R module M, $J(R^G)M$ is properly contained in M [5, Lemma 28.3, p. 314]. Since $J(R^G)$ is central in R, $RJ(R^G)M$ is properly contained in M. Thus $RJ(R^G)$ is T-nilpotent. So $0 \to RJ(R^G) \to J(R) \to J(R)/RJ(R^G) \to 0$ is exact with T-nilpotent first term and nilpotent last term which forces $J(R)$ to be T-nilpotent.

Now $R/J(R)$ is Artinian follows from Theorem 2.3.

We conclude with a question:

Question 2.4. Let R^G be your favorite form of semiperfect ring (for example, QF-3). What is implied about R?

Added in Proof. (1) Question 1.1 has been answered in the negative by W. S. Martindale, 3rd; however, he shows that $|G|J(R^G) \subseteq J(R)$ and hence if R is semisimple with no $|G|$-torsion, then R^G is semisimple.

(2) Theorem 2.2 is also valid if "Krull dimension" is replaced by "Gabriel dimension."

References

[1] G. M. Bergman and I. M. Isaacs, "Rings with fixed-point-free group actions," Proc. London Math. Soc., 27 (1973), 69-87.

[2] M. Cohen and S. Montgomery, "Semisimple Artinian rings of fixed points," Canad. Math. Bull., 18 (1975), 189-190.

[3] J. W. Fisher and J. Osterburg, "Groups acting on Goldie and semiprimary rings," Notices Amer. Math. Soc., 23 (1976), A-75.

[4] ________, "Semiprime ideals in rings with finite group action," to appear.

[5] F. W. Anderson and K. R. Fuller, "Rings and Categories of Modules," Springer-Verlag, Inc., New York, 1973.

[6] R. Gordon and J. C. Robson, "Krull dimension," Memoirs Amer. Math. Soc., 133 (1973).

[7] I. N. Herstein and L. Neumann, "Centralizers in Rings," Annali di Mathematica servie Quarta Tomo CIF (1975), 37-44.

[8] G. Higman, "Groups and rings which have automorphisms with no non-trivial fixed elements," J. London Math. Soc., 32 (1957), 321-334.

[9] J. Lambek, "Lectures on Rings and Modules," Blaisdell, Waltham, Mass., 1966.

[10] J. Levitzki, "On automorphisms of certain rings," Annals of Mathematics, 36 (1935), 984-992.

[11] N. Jacobson, "A note on automorphisms and derivations of Lie algebras," Proc. Amer. Math. Soc., 6 (1955), 281-283.

[12] S. Jondrup, "Groups acting on rings," J. London Math. Soc., (2), 8 (1974), 483-486.

[13] S. Montgomery, "The Jacobson radical and fixed rings of automorphisms," to appear in Communications in Algebra.

[14] D. Passman, "Infinite Group Rings," Marcel Dekker, Inc., New York, 1971.

[15] T. Sundstrom, "Groups of automorphisms of simple rings," J. Algebra, 29 (1974), 555-556.

[16] J. Thompson, "Finite groups with fixed point free automorphisms of prime order," Proc. National Academy Sci. U.S.A., 45 (1959). 578-581.

[17] R. Ware, "Endmorphism rings of projective modules," Trans. of Amer. Math. Soc., 155 (1971), 233-256.

A STUDY OF COTORSION THEORY

V. K. Goel

Ohio University
Athens, Ohio

The purpose of this paper is to study the cotorsion theory initiated by Beachy [2]. A preradical ρ of the category $\mathcal{M}_R$ of right modules over a ring R is called an epiradical if every epimorphism $f:M \to N$ induces an epimorphism $\rho(f):\rho(M) \to \rho(N)$. If ρ is an epiradical and $\rho^2 = \rho$, then ρ is called a cotorsion radical. If M is any R-module with a projective cover P(M), then a functor $\tau^{P(M)}$ given by $\tau^{P(M)}(N) = tr_N(P(M))$, the sum in N of all R-homomorphic images of P(M), is a cotorsion radical. It is shown that there is a 1-1 correspondence between the cotorsion radicals of $\mathcal{M}_R$ and the family $\mathcal{F}$ of right ideals of R satisfying: (i) if $A \in \mathcal{F}$ and $B \subset A$, then $B \in \mathcal{F}$, (ii) if $A_i \in \mathcal{F}$, then $\sum_{i\in I} A_i \in \mathcal{F}$ for any index set I, (iii) if $A \in \mathcal{F}$ and $r \in R$, then $rA \in \mathcal{F}$, and (iv) $\sum_{A\in\mathcal{F}} A$ is an idempotent ideal of R. A cotorsion radical ρ is called a coprime if there exists a nonzero module S having a projective cover P(S) such that $\rho = \tau^{P(S)}$ and $\rho(S') = 0$ for all $S' \subsetneq S$. It is shown that if ρ is a coprime and S is a cosupporting module for ρ (i.e. $\rho(S) = S$ and $\rho(S') = 0$ for all $S' \subsetneq S$) having a projective cover P(S), then $\rho = \tau^{P(S)}$. Also a coprime ρ has a unique (up to isomorphism) cosupporting module having a projective cover which is at the same time ρ-projective. A preradical σ of $\mathcal{M}_R$ is called a torsion radical if for every $M \in \mathcal{M}_R$, $\sigma(L) = \sigma(M) \cap L$, for all $L \subset M$.

In the last section it is shown that if γ is a splitting exact radical then the following are equivalent: (i) every γ-torsion module is projective, (ii) every R-module is γ-injective, (iii) $\gamma(R)$ is semisimple artinian. Throughout all rings have identity and all modules are right unitary.

1. Cotorsion Radicals

Proposition 1.1. Let ρ be a preradical of M_R. Then the following are equivalent:

(i) Every epimorphism $f:M \to N$ induces an epimorphism $\rho(f):\rho(M) \to \rho(N)$.

(ii) $\rho(M) = M\cdot\rho(R)$ for all $M \in M_R$.

(iii) $\rho(M/N) = (\rho(M) + N)/N$ for all $M, N \in M_R$ such that $N \subset M$.

(iv) ρ is a radical and every factor module of a ρ-torsionfree module is ρ-torsionfree.

Proof. (i) $\Rightarrow$ (ii). See Beachy ([2], Proposition 1.3).

(ii) $\Rightarrow$ (iii). Let $M \in M_R$ and $N \subset M$. From (ii), $\rho(M/N) = (M/N)\cdot\rho(R) = (M\cdot\rho(R) + N)/N = (\rho(M) + N)/N$.

(iii) $\Rightarrow$ (iv). Let $M \in M_R$. From (iii), $\rho(M/\rho(M)) = (\rho(M) + \rho(M))/\rho(M) = \rho(M)/\rho(M) = \overline{0}$. Let $\rho(M) = 0$ and $N \subset M$. From (iii) $\rho(M/N) = (\rho(M) + N)/N = N/N = \overline{0}$.

(iv) $\Rightarrow$ (i). See Beachy ([2], Proposition 1.3).

Definition. A preradical ρ satisfying any one of the above conditions is called an epiradical.

Remark 1.2. In the above proposition $\rho(R)$ is an ideal. Also for every ideal A of R the functor ρ defined by $\rho(M) = M\cdot A$, $M \in M_R$ is an epiradical.

Proposition 1.3. There is a one-to-one correspondence between the epiradicals of M_R and the ideals of R.

Proof. Obvious.

Definition. An idempotent epiradical ρ is called a cotorsion radical.

Proposition 1.4. There is a one-to-one correspondence between the cotorsion radicals of M_R and the idempotent ideals of R.

Proof. See Beachy ([2], Theorem 1.4).

Proposition 1.5. For every epiradical ρ there exists a maximal cotorsion radical $\hat{\rho}$ smaller than ρ.

Proof. Let $\mathcal{A} = \{A | A \text{ is an idempotent ideal contained in } \rho(R)\}$. Clearly $\mathcal{A}$ is a partially ordered set under the set theoretic inclusion. By Zorn's lemma let A be a maximal element of $\mathcal{A}$. Define $\hat{\rho}(M) = M \cdot A$, for all $M \in M_R$. Then $\hat{\rho}$ is a cotorsion radical such that $\hat{\rho} \leq \rho$. Suppose there exists a cotorsion radical ρ' such that $\hat{\rho} \leq \rho' \lneq \rho$. Since $\rho'(R)$ is an idempotent ideal properly contained in $\rho(R)$ and $A \subset \rho'(R)$, $A = \rho'(R)$ by maximality of A. Thus $\hat{\rho} - \rho'$.

Proposition 1.6. Let ρ be an epiradical. Let $0 \to A \to B \to C \to 0$ be an exact sequence of R-modules such that A and C are ρ-torsion. Then B is also ρ-torsion.

Proof. Since $C \approx B/A$, $B/A = \rho(B/A) = (\rho(B) + A)/A$. Thus $B = \rho(B) + A$. Since $A \subset B$, $A = \rho(A) \subset \rho(B)$. Hence $\rho(B) = B$.

Proposition 1.7. Let ρ be an epiradical. The following are equivalent:

(i) ρ is idempotent.

(ii) If $0 \to A \to B \to C \to 0$ is an exact sequence such that A and C are ρ-torsionfree, then B is ρ-torsionfree.

Proof. (i) $\Rightarrow$ (ii). Since $C \approx B/A$, we have $\overline{0} = \rho(B/A) = (\rho(B) + A)/A$. Thus $\rho(B) \subset A$. This gives $\rho(B) = \rho(\rho(B)) \subset \rho(A) = 0$.

(ii) $\Rightarrow$ (i). Let A be any R-module. Consider an exact sequence $0 \to \rho(A)/\rho(\rho(A)) \to A/\rho(\rho(A)) \to A/\rho(A) \to 0$. Since ρ is a radical, $\rho(A)/\rho(\rho(A))$ and $A/\rho(A)$ are ρ-torsionfree. Hence from (ii) $A/\rho(\rho(A))$ is also ρ-torsionfree. Thus $\overline{0} = \rho(A/\rho(\rho(A))) = (\rho(A) + \rho(\rho(A)))/\rho(\rho(A))$. Therefore $\rho(A) \subset \rho(\rho(A)) \subset \rho(A)$. Hence $\rho(A) = \rho^2(A)$.

Proposition 1.8. Let F be a family of right ideals satisfying the following three conditions:

(1) If $A \in F$ and $B \subset A$, then $B \in F$.

(2) If $A_i \in F$, then $\sum_{i \in I} A_i \in F$ for any index set I.

(3) If $A \in F$ and $r \in R$, then $rA \in F$.

For every $M \in M_R$, let $\rho(M) = \sum_{A \in F} MA$. Then ρ is an epiradical.

Proof. By definition $\rho(R) = \sum_{A \in F} RA$. Clearly $\sum_{A \in F} A \subset \sum_{A \in F} RA$. Let $x \in \sum_{A \in F} RA$. Then $x = \sum_{i=1}^{n} r_i a_i$, $r_i \in R$ and $a_i \in A_i \in F$. Since from (3) $r_i A_i \in F$, $x \in \sum_{A \in F} A$. Hence $\rho(R) = \sum_{A \in F} A$. Also $\sum_{A \in F} MA = M \cdot \sum_{A \in F} A$. Thus $\rho(M) = M \cdot \rho(R)$ for all $M \in M_R$. Hence, from proposition 1.1, ρ is an epiradical.

Proposition 1.9. Let ρ be an epiradical. Let $F^{\rho} = \{A | \rho(R/A) = \rho(R)/A\}$. Then F^{ρ} satisfies the conditions (1), (2) and (3) stated in the above proposition.

Proof. Note that $F^{\rho} = \{A | A$ is a right ideal of R contained in $\rho(R)\}$.

Theorem 1.10. There is a one-to-one correspondence between the epiradicals and the family of right ideals satisfying the conditions (1), (2) and (3).

Proof. Let F be a family of right ideals satisfying (1), (2) and (3). Define $\rho(M) = M\cdot \sum_{A\in F} A$, for all $M \in M_R$. From proposition 1.8, ρ is an epiradical. From proposition 1.9, $F^{\rho} = \{B | B \subset \sum_{A\in F} A\}$. Let $B \in F^{\rho}$. Then $B \subset \sum_{A\in F} A$. From conditions (1) and (2), $B \in F$. Thus $F^{\rho} \subset F$. Let $A \in F$. Then $A \subset \sum_{A\in F} A$. Thus $A \in F^{\rho}$ and hence $F \subset F^{\rho}$, proving $F = F^{\rho}$.

Remark 1.11. Let F be a family of right ideals satisfying (1), (2) and (3) stated in the proposition 1.8, and the condition that

(4) $\sum_{A\in F} A$ is an idempotent ideal.

Then the corresponding epiradical ρ where $\rho(M) = M\cdot \sum_{A\in F} A$ is a cotorsion radical. Conversely, if ρ is a cotorsion radical, then the corresponding family $F^{\rho} = \{A | \rho(R/A) = \rho(R)/A\}$ satisfies (1), (2), (3) and (4).

Theorem 1.12. There is a one-to-one correspondence between the cotorsion radicals and the family of right ideals satisfying the conditions (1), (2), (3) and (4).

Proof. Follows from theorem 1.10 and the above remark 1.11.

Definition. Let ρ be an epiradical. An R-module P is called ρ-projective if for every epimorphism $f:N \to M$ with $\rho(\ker f) = 0$ and for every homomorphism $g:P \to M$ there exists an R-homomorphism $h:P \to N$ such that $f \circ h = g$.

If in addition the homomorphism $h:P \to N$ is unique then P is called cofaithfully ρ-projective.

Proposition 1.13. The following are equivalent:

(i) P is cofaithfully ρ-projective.

(ii) P is ρ-projective and $\rho(P) = P$.

Proof. (i) $\Rightarrow$ (ii). By the definition of cofaithfully ρ-projectivity P is ρ-projective. Since ρ is a radical, $P/\rho(P)$ is ρ-torsionfree. Thus by (1) there is a unique homomorphism from P to $P/\rho(P)$ and hence $P/\rho(P)$ must be zero.

(ii) $\Rightarrow$ (i). Consider the diagram

$$\begin{array}{ccccc} & & P & & \\ & \overset{h}{\swarrow} & \downarrow g & & \\ N & \xrightarrow{f} & M & \rightarrow & 0 \end{array}$$

such that $\ker f$ is ρ-torsionfree. Since P is ρ-projective, there exists $h:P \rightarrow N$ such that $foh = g$. Suppose there exists $h':P \rightarrow N$ such that $foh' = g$. Then $foh = g = foh'$. Define $(h-h'):P \rightarrow N$ by $(h-h')(p) = hp - h'p$. Then $(h-h')P \subset \ker f$. Since P is ρ-torsion, $\ker f$ is ρ-torsionfree and ρ is a preradical, we get $(h-h')P = 0$. Hence $h = h'$.

After having proved the following propositions 1.14 to 1.18 we have found that they are also contained in Bland [6] with a slightly different terminology. For the sake of completeness we shall prove propositions 1.15 and 1.18 which are needed in further development of our theory.

Proposition 1.14. Let ρ be an epiradical and $0 \rightarrow A \rightarrow B \rightarrow C \rightarrow 0$ be an exact sequence in which A is ρ-torsion and B is ρ-projective. Then C is ρ-projective.

Proof. See Bland ([6], Theorem 2.7).

Proposition 1.15. Let ρ be an epiradical and $0 \rightarrow A \overset{i}{\rightarrow} B \overset{t}{\rightarrow} C \rightarrow 0$ be an exact sequence in which A is ρ-torsionfree and C is ρ-projective. Then the sequence splits. If in addition B is ρ-torsion (or A is small in B), then $B \approx C$.

Proof. Consider the diagram

$$\begin{array}{ccc} & C & \\ & \downarrow i & \\ B & \xrightarrow{t} C & \rightarrow 0 \end{array}$$

Since C is ρ-projective and $\rho(\ker t) = 0$, the sequence $0 \to A \xrightarrow{i} B \xrightarrow{t} C \to 0$ splits and hence $B = A \oplus H$ where $H \approx C$. Assume $\rho(B) = B$, then $B = \rho(A \oplus H) \subset \rho(H) \subset H \subset B$. Hence $H = B$ proving $B \approx C$. If A is small in B, then $B = A \oplus H$ which gives $B = H$. Thus $B \approx C$.

Proposition 1.16. Let ρ be an epiradical and $0 \to A \to B \to C \to 0$ be an exact sequence in which C is ρ-projective and A is small in B. Then A is ρ-torsion.

Proof. See Bland ([6], Theorem 2.7).

Remark 1.17. Let ρ be an epiradical. If A is a small submodule of B with B/A ρ-torsion, then B is ρ-torsion.

Proposition 1.18. Let ρ be a cotorsion radical and M be an R-module having a projective cover. Then there exists a ρ-projective module P (which is unique up to isomorphism) and an epimorphism $f:P \to M$ such that $\ker f$ is small in P and $\rho(\ker f) = 0$. If in addition M is ρ-torsion, then P is ρ-torsion.

Proof. Let $g:P(M) \to M \to 0$ be a projective cover of M. Consider an exact sequence $0 \to \rho(\ker g) \to P(M) \to P(M)/\rho(\ker g) \to 0$. Then from proposition 1.14, $P(M)/\rho(\ker g)$ is ρ-projective. Define $f:P(M)/\rho(\ker g) \to M$ by $f(\overline{x}) = g(x)$, $\overline{x} \in P(M)/\rho(\ker g)$. Then $\ker f = \ker g/\rho(\ker g)$ is small in $P(M)/\rho(\ker g)$ and $\rho(\ker f) = 0$. Thus $P = P(M)/\rho(\ker g)$ is a ρ-projective module with an epimorphism $f:P \to M$ such that $\ker f$ is small in P and $\rho(\ker f) = 0$.

Suppose there exists P' a ρ-projective module and an epimorphism $f':P' \to M$ such that $\ker f'$ is small in P' and $\rho(\ker f') = 0$. Consider the diagram

$$\begin{array}{ccccc} & & P & & \\ & \overset{h}{\swarrow} & \downarrow f & & \\ P' & \underset{f'}{\to} & M & \to & 0 \end{array}$$

Since P is ρ-projective there exists a homomorphism $h:P \to P'$ such that $f' \circ h = f$. Clearly, $h(P) + \ker f' = P'$, and therefore $h(P) = P'$. Thus $0 \to \ker h \to P \to P' \to 0$ is an exact sequence. Since $f' \circ h = f$, $\ker h$ is small in P. Hence as in the proof of proposition 1.15, $\ker h = 0$. Thus $P \approx P'$.

Assume M is ρ-torsion with an epimorphism $f:P \to M$ such that P is ρ-projective, $\ker f$ is small in P, and $\rho(\ker f) = 0$. Since $P/\ker f$ is ρ-torsion, it follows from remark 1.17 that P is ρ-torsion.

Lemma 1.19. Let ρ be an epiradical. If A is ρ-torsion then every ρ-torsionfree submodule of A is small in A.

Proof. Let $B \subset A$ be such that $\rho(B) = 0$. Let $B + C = A$ and $a \in A = A \cdot \rho(R)$. Then $a = \sum_{i=1}^{n} a_i r_i$, $a_i \in A$ and $r_i \in \rho(R)$. Also $a_i = b_i + c_i$, $b_i \in B$ and $c_i \in C$. Since $B \cdot \rho(R) = 0$, $a = \sum_{i=1}^{n} (b_i + c_i) r_i = \sum_{i=1}^{n} c_i r_i$. Hence $A = C$ completing the proof.

Next we consider the associated cotorsion radical of a module. Let M,N be two R-modules. Denote by $\mathrm{rad}^M(N) = \mathrm{tr}_N(M)$ the sum in N of all R-homomorphic images of M. The functor rad^M is an idempotent preradical. If P is a projective module we shall denote rad^P by τ^P.

Proposition 1.20. If P is a projective module, then τ^P is a cotorsion radical.

Proof. See Beachy ([2], Proposition 1.5).

Definition. Let M be an R-module having a projective cover P(M). We shall call $\tau^{P(M)}$ the associated cotorsion radical of M whenever the projective cover P(M) of M exists.

Proposition 1.21. If $\{\rho_i | i \in I\}$ is a set of epiradicals (cotorsion radicals) then $\rho = \sum_{i \in I} \rho_i$ defined as $\rho(M) = M \cdot (\sum_{i \in I} \rho_i(R))$ for all $M \in M_R$ is an epiradical (cotorsion radical).

Proof. Straightforward.

Lemma 1.22. Let M,N be two R-modules having their projective covers P(M) and P(N) respectively. Then $\tau^{P(M \oplus N)} = \tau^{P(M)} + \tau^{P(N)}$.

Proof. Observe the fact that $P(M \oplus N) = P(M) \oplus P(N)$.

Lemma 1.23. If $f: M \to N$ is an epimorphism, then $\text{rad}^M \geq \text{rad}^N$.

Proof. Obvious.

Lemma 1.24. Let M be an R-module having a projective cover P(M). Then $\tau^{P(M)}(M) = M$, and for every cotorsion radical ρ for which M is ρ-torsion P(M) is ρ-torsion and $\tau^{P(M)} \leq \rho$.

Proof. Let $f: P(M) \to M$ be an epimorphism such that ker f is small in P(M). Let ρ be a cotorsion radical with $\rho(M) = M$. Since ρ is a cotorsion radical $\rho(f): \rho(P(M)) \to M$ is an epimorphism and $\rho(P(M)) = P(M)$. For any $N \in M_R$, $\tau^{P(M)}(N) = tr_N(P(M))$. Since P(M) is ρ-torsion for each homomorphism $h: P(M) \to N$ we have $h(P(M)) \subset \rho(N)$. Hence $\tau^{P(M)}(N) \subset \rho(N)$.

Remark 1.25. If $f: M \to N$ is an epimorphism and P(M) and P(N) are projective covers of M and N respectively, then $\tau^{P(M)} \geq \tau^{P(N)}$.

Proposition 1.26. Let M be an R-module having a projective cover P(M). Then rad^M is a cotorsion radical iff $\tau^{P(M)} = \text{rad}^M$.

Proof. Let $f: P(M) \to M \to 0$ be a projective cover of M. From lemma 1.23,

$\tau^{P(M)} \geq rad^M$. Assume rad^M is a cotorsion radical. Since $rad^M(M) = M$, from lemma 1.24, $\tau^{P(M)} \leq rad^M$. Thus $\tau^{P(M)} = rad^M$. Converse is obvious.

Proposition 1.27. If ρ is a cotorsion radical of M_R, then $\rho = rad^{\rho(R)}$.

Proof. See Beachy ([2], Proposition 1.6).

The following theorem is due to Beachy [2]. Here we provide another proof.

Theorem 1.28. If every idempotent ideal of R has a projective cover as a right R-module, then every cotorsion radical ρ of M_R is of the form $\tau^{P(\rho(R))}$.

Proof. From proposition 1.27, $\rho = rad^{\rho(R)}$. Since $\rho(R)$ is an idempotent ideal it has a projective cover $P(\rho(R))$. Thus from proposition 1.26, $\rho = \tau^{P(\rho(R))}$.

Theorem 1.29. Let M,N be two R-modules with projective covers P(M) and P(N) respectively. Let $f: M \to N \to 0$ be an epimorphism with ker f small in M. Then $\tau^{P(M)} = \tau^{P(N)}$.

Proof. From ([1], lemma 27.5), $P(M) \approx P(N)$. Thus $\tau^{P(M)} = \tau^{P(N)}$.

2. Coprimes

Definition. Let ρ be an epiradical. A nonzero R-module S is called a cosupporting module for ρ if

(i) $\rho(S) = S$

(ii) $\rho(S') = 0$ for all $S' \subsetneq S$.

Proposition 2.1. Let ρ be a cotorsion radical and S be a cosupporting module for ρ. Then we have the following:

(1) Every proper submodule S' of S is small in S.

(2) Every nonzero factor module of S is also a cosupporting module for ρ.

(3) If $f:T \to S$ is an epimorphism with $\rho(T) = T$ and $\rho(\ker f) = 0$, then T is also a cosupporting module for ρ.

(4) If $P(S)$ is a projective cover for S, then S is cosupporting for $\tau^{P(S)}$.

Proof. (1) Let $S' \subsetneq S$ and $S' + A = S$, for some submodule A of S. If $A \subsetneq S$, then $\rho(A) = 0$. Since S' is also ρ-torsionfree, we get $\rho(S) = 0$, a contradiction. Hence $A = S$ and S' is small in S.

(2) Let S/S' be a nonzero factor module of S. Clearly, $\rho(S/S') = S/S'$. Let $S''/S' \subsetneq S/S'$. Then $S'' \subsetneq S$ and hence $\rho(S'') = 0$. Since ρ is a cotorsion radical $\rho(S''/S') = 0$. Thus S/S' is also a cosupporting module for ρ.

(3) Let T be an R-module and $f:T \to S$ be an epimorphism with $\rho(T) = T$ and $\rho(\ker f) = 0$. From lemma 1.19, $\ker f$ is small in T. Let $T' \subsetneq T$. If $T' \subset \ker f$, then $\rho(T') = 0$. Assume $T' \not\subset \ker f$. Consider an exact sequence $0 \to \ker f \to T' + \ker f \to f(T') \to 0$. If $f(T') = S$, then the smallness of $\ker f$ in T gives $T' = T$, a contradiction. Hence $f(T') \subsetneq S$ and therefore $\rho(f(T')) = 0$. Thus from proposition 1.7, $\rho(T' + \ker f) = 0$. Hence $\rho(T') = 0$ proving T is a cosupporting module for ρ.

(4) Clearly $\tau^{P(S)}(S) = S$. Since $\rho(S) = S$ from lemma 1.24, $\tau^{P(S)} \leq \rho$. Thus for all $S' \subsetneq S$, $\tau^{P(S)}(S') \subset \rho(S') = 0$. Hence S is a cosupporting module for $\tau^{P(S)}$.

Lemma 2.2. Let S be a cosupporting module for an epiradical ρ and A be a ρ-torsion module. Then every nonzero homomorphism $f:A \to S$ is an epimorphism.

Proof. Assume $f(A) \subsetneq S$. Then $\rho(f(A)) = 0$. Therefore $f(A) = 0$, a contradiction. Hence f is an epimorphism.

Definition. A cotorsion radical ρ is called a coprime if there exists a cosupporting module S for ρ having a projective cover $P(S)$ such that $\rho = \tau^{P(S)}$.

If S is a sample R-module with a projective cover $P(S)$ and $\rho = \tau^{P(S)}$, then obviously S is a cosupporting module for ρ and hence ρ is a coprime. Thus in a semiperfect ring every maximal right ideal of R gives rise to a coprime.

Lemma 2.3. Let $f:M \to N$ be an epimorphism with $\ker f$ small in M. If each proper submodule of N is small in N, then each proper submodule of M is small in M.

Proof. Obvious.

Lemma 2.4. Let S and S' be non-isomorphic simple R-modules and S has a projective cover $P(S)$. Then $\tau^{P(S)}(S') = 0$.

Proof. Let $f:P(S) \to S$ be an epimorphism with $\ker f$ small in $P(S)$. Let N be any submodule of P such that $N \not\subset \ker f$. Then $0 \neq f(N + \ker f) = S$ and therefore $N = P(S)$. Hence $\ker f$ is a unique maximal submodule of $P(S)$. Suppose $\tau^{P(S)}(S') \neq 0$, then $\tau^{P(S)}(S') = S'$. Thus there exists a nonzero homomorphism $h:P(S) \to S'$. Since S is simple, it follows from lemma 2.3, that each proper submodule of $P(S)$ is small in $P(S)$. Thus $\ker h$ is small in $P(S)$. Since S' is also simple, $\ker h$ is also a unique maximal submodule of $P(S)$. Hence $\ker h = \ker f$ and therefore $S \approx S'$ a contradiction. Thus $\tau^{P(S)}(S') = 0$.

Remark 2.5. If R is semiperfect then the coprimes which arise from simple modules are in one-to-one correspondence with the isomorphism classes of the simple modules.

Proposition 2.6. If R is semiperfect, then all simple modules are ismorphic iff 1(the identity cotorsion radical) is a coprime.

Proof. Suppose all simple modules are ismorphic. Let S be a simple R-module. Then clearly S is a cosupporting module for 1. Suppose $1 \neq \tau^{P(S)}$, then $\tau^{P(S)}(R) \subsetneq R$. Thus there exists a proper maximal right ideal A containing $\tau^{P(S)}(R)$ and hence $\tau^{P(S)}(R/A) = \overline{0}$. Since $R/A \approx S$, $\tau^{P(S)}(R/A) = R/A$, a contradiction. Thus $1 = \tau^{P(S)}$. Hence 1 is a coprime.

Conversely, let 1 be a coprime. Therefore $1 = \tau^{P(S)}$ where S is a cosupporting module for 1. Clearly S is a simple module. Let S' be any other simple module. Suppose $S \not\approx S'$, then from lemma 2.4, $\tau^{P(S)}(S') = 0$. Thus $S' = 0$, a contradiction. Hence all simple modules are isomorphic.

Proposition 2.7. Let ρ be a coprime and S be a cosupporting module for ρ having a projective cover P(S). Then $\rho = \tau^{P(S)}$.

Proof. Since ρ is a coprime there exists a cosupporting module T for ρ with a projective cover P(T) such that $\rho = \tau^{P(T)}$. Let $f: P(T) \to T$ be an epimorphism with ker f small in P(T). Since $\tau^{P(T)}(S) = S$, there exists a nonzero homomorphism $g: P(T) \to S$ which is an epimorphism from lemma 2.2. Since T is a cosupporting module for ρ, it follows from lemma 2.3 that every proper submodule of P(T) is small in P(T). Thus ker g is small in P(T). Hence from theorem 1.29, $\rho = \tau^{P(T)} = \tau^{P(S)}$.

Theorem 2.8. Let ρ be a coprime. Then, up to isomorphism, there is only one cosupporting module for ρ having a projective cover which is at the same time ρ-projective.

<u>Proof</u>. Let ρ be a coprime. Therefore, there exists a cosupporting module S for ρ having a projective cover $P(S)$ such that $\rho = \tau^{P(S)}$. Let $g:P(S) \to S$ be an epimorphism with $\ker g$ small in $P(S)$. Then from proposition 1.18, $P(S)/\rho(\ker g)$ is ρ-projective and $f:P(S)/\rho(\ker g) \to S$ defined by $f(\bar{x}) = g(x)$, $\bar{x} \in P(S)/\rho(\ker g)$ is an epimorphism with $\rho(\ker f) = 0$. Since S is cosupporting for ρ, it follows from proposition 2.1, $P(S)/\rho(\ker g)$ is also a cosupporting module for ρ. Also $P(S)$ is a projective cover of $P(S)/\rho(\ker g)$. Thus the existance of at least one such cosupporting module for ρ having a projective cover which is at the same time ρ-projective is shown.

So let U and V be ρ-projective cosupporting modules for ρ with their projective cover $P(U)$ and $P(V)$ respectively. Let $f:P(U) \to U$ be an epimorphism with $\ker f$ small in $P(U)$. Since ρ is a coprime and U is a cosupporting module for ρ, $\rho = \tau^{P(U)}$. Also V being the cosupporting module for ρ we have $\rho(V) = V$. Therefore, there exists a homomorphism $h:P(U) \to V$ which is an epimorphism from lemma 2.2. Define $t:V \to U/f(\ker h)$ by $t(v) = \overline{f(x)}$, where $v = h(x)$ for some $x \in P(U)$. Then t is well defined epimorphism. Consider the diagram

$$\begin{array}{ccccc} & & U & & \\ & \swarrow^{s} & \downarrow \pi & & \\ V & \xrightarrow{t} & U/f(\ker h) & \to & 0 \end{array}$$

Since V is cosupporting for ρ, $\ker t$ is ρ-torsionfree. Hence by ρ-projectivity of U there exists a homomorphism $s:U \to V$ such that $t \circ s = \pi$. Again from lemma 2.2, s is an epimorphism. So consider an exact sequence $0 \to \ker s \to U \to V \to 0$. Since $\rho(\ker s) = 0$, V is ρ-projective and $\ker s$ is small in U, from proposition 1.15, $U \approx V$.

Definition. Let M be a nonzero R-module having a projective cover P(M). Then M is called costable if $\rho(M)$ is small in M for every cotorsion radical ρ which is properly smaller than $\tau^{P(M)}$.

Obviously, if M has a projective cover P(M) and all proper submodules of M are small in M, then M is costable.

Proposition 2.9. Let M be an R-module having a projective cover for all of its factor modules. Then the following are equivalent:

(i) M is costable.

(ii) If $N \subsetneq M$, then $\tau^{P(M/N)} = \tau^{P(M)}$

Proof. (i) $\Rightarrow$ (ii) Let $N \subsetneq M$. Since $\tau^{P(M)}(M) = M$, $\tau^{P(M)}(M/N) = M/N$. Hence $\tau^{P(M/N)} \leq \tau^{P(M)}$. Assume $\tau^{P(M/N)} \neq \tau^{P(M)}$, then $\tau^{P(M/N)} \lneq \tau^{P(M)}$. Thus from (i) $\tau^{P(M/N)}(M)$ is small in M. Also $\tau^{P(M/N)}(M/N) = M/N$ gives $M = \tau^{P(M/N)}(M) + N$. Hence $M = N$, a contradiction. Thus $\tau^{P(M/N)} = \tau^{P(M)}$.

(ii) $\Rightarrow$ (i) Let ρ be a cotorsion radical such that $\rho \lneq \tau^{P(M)}$. Let $\rho(M) + K = M$. Then $\rho(M/K) = M/K$ and therefore $\tau^{P(M/K)} \leq \rho$. Assume $K \subsetneq M$, then (ii) gives $\tau^{P(M/K)} = \tau^{P(M)}$, a contradiction. Thus $K = M$ proving M is costable.

3. Exact Radicals

Definition. Let γ be a preradical of M_R. γ is called an exact radical if for every exact sequence $0 \to A \to B \to C \to 0$ we have an exact sequence $0 \to \gamma(A) \to \gamma(B) \to \gamma(C) \to 0$.

Clearly γ is an exact radical iff γ is a torsion radical as well as a cotorsion radical.

Lemma 3.1. Let R be ring and A be a right ideal of R such that for every $x \in A$, there exists $a_x \in A$ such that $x = xa_x$. Then for every finite set $\{x_1, x_2, \ldots, x_n\}$ of elements in A, there exists a ε A such that $x_i = x_i a$.

Proof. It is clearly sufficient to prove the assertion for two elements x_1, x_2 of A. Since $x_1 = x_1 a_1$, for some $a_1 = a_{x_1} \varepsilon A$ and $x_2 - x_2 a_1 = (x_2 - x_2 a_1) a_2$, for some $a_2 = a_{(x_2 - x_2 a_1)} \varepsilon A$, we get $x_1 = x_1 a$ and $x_2 = x_2 a$ where $a = (a_1 + a_2 - a_1 a_2)$.

Proposition 3.2. Let ρ be a cotorsion radical. Then the following are equivalent:

(1) $\rho(N) = N \cap \rho(M)$, for all $N \varepsilon M$.

(2) $\rho(A) = A \cap \rho(R)$, for all right ideals A in F^{ρ}.

(3) $A = A \cdot \rho(R)$, for all right ideals A in F^{ρ}.

(4) $R/\rho(R)$ is flat as left R-module.

Proof. (1) $\Rightarrow$ (2) and (2) $\Rightarrow$ (3) are obvious.

(3) $\Rightarrow$ (1) Let $N \subset M$ and $x \varepsilon N \cap \rho(M)$. Then $x = \sum_{i=1}^{n} m_i r_i$ where $x \varepsilon N$, $m_i \varepsilon M$ and $r_i \varepsilon \rho(R)$. From (3) $rR = rR \cdot \rho(R)$, for all $r \varepsilon \rho(R)$. Therefore $r = ry_r$ for some $y_r \varepsilon \rho(R)$. Hence from lemma 3.1, there exists $y \varepsilon \rho(R)$ such that $r_i = r_i y$. Thus $xy = x$ and hence $x \varepsilon \rho(N)$ proving $\rho(N) = N \cap \rho(M)$.

(1) $\Leftrightarrow$ (4) See Ramamurthi ([8], Proposition 1.1).

Corollary 3.3. A cotorsion radical ρ is an exact radical iff A is ρ-torsion for all right ideals A in F^{ρ}.

Recall if σ is a torsion radical, then the collection of right ideals A such that $\sigma(R/A) = R/A$ is denoted by F_{σ}.

Proposition 3.4. Let σ be a torsion radical. Then the following are equivalent:

(1) $\sigma(M) = M \cdot \sigma(R)$ for all $M \varepsilon M_R$.

(2) $\sigma(R/A) = (\sigma(R) + A)/A$ for all $A \varepsilon F_{\sigma}$.

(3) $A + \rho(R) = R$ for all $A \in F_\sigma$.

Proof. (1) ⇒ (2) and (2) ⇒ (3) are obvious.

(3) ⇒ (1) Let $x \in M\cdot\sigma(R)$. Then $x = \sum_{i=1}^{n} m_i r_i$ where $m_i \in M$ and $r_i \in \sigma(R)$. Since $r_i \in \sigma(R)$, there exists $B_i \in F_\sigma$ such that $r_i B_i = 0$. Let $B = \bigcap_{i=1}^{n} B_i$, then $B \in F_\sigma$ and $xB = 0$. Hence $x \in \sigma(M)$. Therefore $M\cdot\sigma(R) \subset \sigma(M)$. Let $x \in \sigma(M)$. Then there exists $A \in F_\sigma$ such that $xA = 0$. From (3), $A + \sigma(R) = R$. Therefore $1 = a + b$ for some $a \in A$, $b \in \sigma(R)$. Hence $x = xb \in M\cdot\sigma(R)$. Thus $\sigma(M) = M\cdot\sigma(R)$.

Corollary 3.5. A torsion radical σ is an exact radical iff $A + \sigma(R) = R$ for all $A \in F_\sigma$.

Next we give a result proved earlier by Ramamurthi ([8], Proposition 1.6) and extend a result of Ramamurthi and Rutter ([9], Corollary 1.4).

Proposition 3.6. If R is a ring, each nonzero ideal of which contains a nonzero divisor, then M_R has only two exact radicals, namely, the zero functor and the identity functor.

Proof. Assume γ is an exact radical such that $\gamma \neq 0$, the zero functor. Then $\gamma(R) \neq 0$. Let x be nonzero divisor in $\gamma(R)$. Since $x \in \gamma(R)$, there exists $A \in F_\gamma$ such that $xA = 0$. Hence $A = 0$ and therefore $0 \in F_\gamma$. Thus $\gamma = 1$, the identity functor.

Proposition 3.7. Let γ be an exact radical. Then the following are equivalent:

(1) The γ-torsion class is closed under direct products.

(2) $\gamma(R)$ is a direct summand of R as a left R-module.

(3) γ commutes with direct products.

Proof. (1) ⇒ (2) Since the γ-torsion class is closed under direct products from ([10], Proposition 6.12), $A = \cap A_i \in F_\gamma$ where $A_i \in F_\gamma$. Then $\gamma(R) = \{r \in R | rA = 0\}$. Hence $\gamma(R) = \ell(A)$, the left annihilator of A. Since γ is a radical and $A \in F_\gamma$, $A + \gamma(R) = R$. Clearly, $\gamma(R)$ has a right identity. Thus $\gamma(R)$ is a direct summand of R as left R-module.

(2) ⇒ (3) See Ramamurthi and Rutter ([9], Proposition 1.3).

(3) ⇒ (1) Obvious.

Theorem 3.8. Let γ be a splitting exact radical. Then the following are equivalent:

(1) Every γ-torsion module is projective.

(2) Every R-module is γ-injective.

(3) $\gamma(R)$ is semisimple artinian.

Proof. (1) ⇒ (2) Let M be any R-module. Consider the diagram

$$\begin{array}{ccc} 0 \to L & \xrightarrow{i} & N \\ f \downarrow & \swarrow g & \\ M & & \end{array}$$

such that $\gamma(N/L) = N/L$. Since $0 \to L \to N \to N/L \to 0$ is an exact sequence and N/L is projective, the sequence splits. Hence there exists $g: N \to M$ such that g is an extension of f on L.

(2) ⇒ (3) Since γ is splitting, γ is centrally splitting by ([9], Theorem 1.5). Hence $R = \gamma(R) \oplus A$ with $1 = e + f$, $e \in \gamma(R)$, $f \in A$ and e, f are central idempotents of R. Let I be a right ideal of $\gamma(R)$. Then I is also a right ideal of R. Since I is γ-injective, there exists an R-homomorphism $f: \gamma(R) \to I$ such that the diagram

$$\begin{array}{ccc} 0 \to I & \xrightarrow{i} & \gamma(R) \\ i \downarrow & \swarrow f & \\ I & & \end{array}$$

commutes. Hence I is a direct summand of $\gamma(R)$ as R-module. Thus I is also a direct summand of $\gamma(R)$ as $\gamma(R)$-module proving $\gamma(R)$ is semi-simple artinian.

(3) ⇒ (1) Let P be a γ-torsion module. Since P is also a $\gamma(R)$-module, P is projective as $\gamma(R)$-module. Consider the diagram

$$\begin{array}{ccccc} & & P & & \\ & \overset{h}{\swarrow} & \downarrow g & & \\ N & \xrightarrow[f]{} & M & \rightarrow & 0 \end{array}$$

where $M,N \in \mathcal{M}_R$, g is R-homomorphism and f is R-epimorphism. Then f,g are also $\gamma(R)$-homomorphism. Thus there exists $h:P \to N$ a $\gamma(R)$ = homomorphism such that $f \circ h = g$. Let $p \in P$, $r \in R$. Since $P = P\cdot\gamma(R)$, $pr = \sum_{i=1}^{n} p_i r_i$ where $p_i \in P$, $r_i \in \gamma(R)$. As e is a central idempotent of R and an identity of $\gamma(R)$, $pre = pr$. Hence $h(pr) = h(pre) = h(p)re = h(p)er = h(pe)r = h(p)r$. Thus h is also an R-homomorphism proving P is projective as R-module.

Acknowledgments

The author is indebted to Professor S. K. Jain for encouragement and advice at all stages of the present work.

The author wishes to express his sincere thanks to the referee for his valuable comments and suggestions.

References

[1] F. Anderson and K. Fuller, "Rings and categories of modules," Springer-Verlag, Graduate text in mathematics, Vol. 13, 1973.

[2] J. Beachy, "Cotorsion radicals and projective modules," Bull. Austal. Math. Soc. 5 (1971), 241-253.

[3] ________, "A generalization of injectivity," Pacific J. Math. 41 (1972), 313-327.

[4] L. Bican, P. Jambor, T. Keppa and P. Němec, "Preradicals," Comment. Math. Univ. Carol. 15 (1974), 75-83.

[5] ________, "Hereditary and cohereditary preradicals," Czech. Math. J. 26 (1976), 192-206.

[6] P. Bland, "Divisible and codivisible modules," Math. Scand. 34 (1974), 153-161.

[7] O. Goldman, "Rings and modules of quotients," J. Algebra 13 (1969), 10-47.

[8] V. Ramamurthi, "On splitting cotorsion radicals," Proc. Amer. Math. Soc. 39 (1973), 457-461.

[9] ________ and E. Rutter, "On cotorsion radicals," Pacific J. Math. 62 (1976), 163-172.

[10] B. Stenström, "Rings of Quotients," Springer-Verlag, Band 217, 1975.

APPLYING GROUP REPRESENTATION TO NONASSOCIATIVE ALGEBRAS

Irvin Roy Hentzel*

Iowa State University
Ames, Iowa

Suppose that we have an associative ring R which satisfies this particular identity:

1) $abcd + cbad - dabc \equiv 0.$

What can be said about the ring? Letting $a = b = c = d = x$, we conclude $x^4 = 0$ for all x in R; some minor restrictions on characteristic and the Nagata-Higman theorem, [4, p. 274], would imply that R is nilpotent of index at most 15. Is it nilpotent of index at most 4? If we substitute a for b and c for d in the identity, and add the two identities together, perhaps all terms will cancel except for those of one particular arrangement. We would then know that the sum of terms each equaling zero is n(abcd) for some integer n. If the characteristic is not n, we would have $abcd = 0$ is an identity, and R is nilpotent of index at most 4. The group ring approach enables us to decide if it is possible to make substitutions of this type that would prove $R^4 = 0$. The group ring approach also tells us which substitutions should be made and which characteristics need to be excluded.

---

*This paper was written while the author held a grant from the Iowa State University Science and Humanities Research Institute.

The above example 1) is a polynomial identity on an associative ring. The technique is not limited to associative rings. It is possible to examine identities similar to 1) in nonassociative rings; there, the order of multiplication must be specified: for example, $(ab)(cd) + (cb)(ad) - (da)(bc) \equiv 0$. This type of identity, where all terms involved are identically associated, rarely occurs. More often one has identities involving different associations. For example, the right alternative law is: $(ab)c + (ac)b - a(bc) - a(cb) \equiv 0$. Most often the identity involves one or more complicated expressions. Rather than consider the expressions as made up of associated products summed together, it is easier to consider one or more functions $f:R^n \to R$ and identities involving these functions. An example is the associator function $(a,b,c) \equiv (ab)c-a(bc)$. With respect to the associator function, the right alternative law is $(a,b,c) + (a,c,b) \equiv 0$. The commutator $[a,b] \equiv ab - ba$ is another useful function; combinations of associators and commutators play an important part in the theory. The function $[a,(b,c,d)]$ really represents a sum of 4 terms.

The group ring approach can be used to prove results about identities when expressed through the use of one or more of these functions. It can be used to prove general results using Young diagrams and known properties of the group ring. When there are few variables (9 or less), one can calculate the matrix representation of the group ring and work the problem using standard matrix techniques such as determinant and row operations.

Let R be a (not necessarily associative) algebra over a field F. Let $f:R^n \to R$ be a function of n arguments. Let S_n be the symmetric group on n objects. By f_π we mean to permute the arguments of f by π before f is evaluated on them. As an example, $f_{(12)}(a,b,c) = f(b,a,c); f_{(123)}(a,b,c) = f(c,a,b)$. We multiply permutations from left to right. $\sigma\pi$ means apply σ first and then apply π. Permutations apply to positions, not elements. If $\pi = (123)$, then $f_\pi(x_1,x_2,x_3) = f(x_3,x_1,x_2)$; $f_\pi(x_1,x_2,x_3) \neq f(x_2,x_3,x_1) = f(x_{1\pi},x_{2\pi},x_{3\pi})$.

If $\Phi = \sum_{\pi \in S_n} \lambda_\pi \pi$ is any element of the group ring on the symmetric group S_n, we associate the function $f_\Phi = \sum_{\pi \in S_n} \lambda_\pi f_\pi$ with Φ. We call Φ an identity based on f if f_Φ is the zero function. If H is an additive subgroup of R (H not necessarily an ideal), and $f_\Phi(R^n) \subset H$, we call Φ an identity mod H of R based on the function f.

The set of all identities of R mod H is a left ideal of the group ring on the symmetric group S_n. If $\Phi_1,\Phi_2,\ldots,\Phi_s$ are known identities of R mod H, the set of all identities implies by them through substitution of arguments and adding or subtracting will be the left ideal of the group ring generated by $\Phi_1,\Phi_2,\ldots,\Phi_s$. Two collections of identities $\Phi_1,\Phi_2,\ldots,\Phi_s$ and $\Phi_1',\Phi_2',\ldots,\Phi_t'$ are equivalent if they generate the same left ideal of the group ring.

Using this techique, identity manipulation can be carried out by group ring structure and by matrix techniques.

Let us return to the associative algebra R over a field F which satisfies the identity mentioned in the introduction: abcd + cbad - dabc. The function f(a,b,c,d) is the product abcd. The identity Φ is the element $\Phi = I + (13) - (1234)$ of the group ring. What is the left ideal of the group ring on S_4 generated by Φ? If the characteristic of the field F is not 2 or 3, the representation of Φ is:

$$\Phi = (1) \oplus (1) \oplus \begin{pmatrix} 1 & 0 \\ 0 & 1 \end{pmatrix} \oplus \begin{pmatrix} 3 & -2 & 0 \\ 1 & 0 & -1 \\ 1 & -1 & 2 \end{pmatrix} \oplus \begin{pmatrix} -1 & 2 & 0 \\ -1 & 2 & 1 \\ -1 & 1 & 0 \end{pmatrix} .$$

Since the matrix in each summand is invertible, Φ is invertible, and the left ideal generated by Φ is the entire group ring. In particular, the identity permutation I is an identity of R based on f. Hence $0 = f_I(a,b,c,d) = abcd$, and R is nilpotent of index at most 4 (characteristic $\neq 2, \neq 3$).

We now consider identities on an algebra which are based on more than one function. Let $f_1, f_2, f_3, \ldots, f_k$ be functions from $R^n \to R$. Let P be the group ring on S_n. We take M to be the direct sum of k copies of P (one copy for each function). M will be a module over P where addition is defined coordinatewise and multiplication is defined by

$$\Phi(\Phi_1 \oplus \Phi_2 \oplus \Phi_3 \oplus \ldots \oplus \Phi_k) = \Phi\Phi_1 \oplus \Phi\Phi_2 \oplus \Phi\Phi_3 \oplus \ldots \oplus \Phi\Phi_k .$$

Next we show how identities of the ring R correspond to submodules of M. We say that $\Phi_1 \oplus \Phi_2 \oplus \ldots \oplus \Phi_k \in M$ is an identity of R based on the functions $f_1 \oplus f_2 \oplus \ldots \oplus f_k$ if the function $(f_1)_{\Phi_1} + (f_2)_{\Phi_2} + \ldots + (f_k)_{\Phi_k}$ takes the value $0 \in R$ for all possible choices of its n arguments. The set of all identities of R based on $f_1 \oplus f_2 \oplus \ldots \oplus f_k$ forms a submodule of M. An identity can be deduced from a collection of identities if it is in the submodule generated by them. Two sets of identities are equivalent if they generate the same submodule.

When working with M, if $\Phi_1 \oplus \Phi_2 \oplus \ldots \oplus \Phi_i \oplus \ldots \oplus \Phi_k$ is an identity and Φ_i is invertible, then $\Phi_i^{-1}\Phi_1 \oplus \Phi_i^{-1}\Phi_2 \oplus \ldots \oplus I \oplus \ldots \oplus \Phi_i^{-1}\Phi_k$ is also an identity. This expresses the ith function as a combination of the remaining functions. If $\Phi\Phi_i = 0$, then

$\Phi\Phi_1 \oplus \Phi\Phi_2 \oplus \ldots \oplus 0 \oplus \ldots \oplus \Phi\Phi_k$ is an identity not involving the ith function. Using these two techniques, the system of identities can be simplified.

If R is a nonassociative algebra over a field of characteristic zero, then any (nonassociative) polynomial identity is equivalent to linear homogeneous polynomial identities. These linear homogeneous polynomials will be sums of products of the same degree but with various associations and arrangements of the elements. For degree n, let a(n) be the number of possible associations of a product of n elements. Let M_n be the direct sum of a(n) copies of the group ring on S_n. The identities satisfied by R will be expressed by a sequence of submodules $L_1, L_2, \ldots, L_n, \ldots$ of the $M_1, M_2, \ldots, M_n, \ldots$.

In characteristic zero, the group ring is semisimple. To tell if a particular identity is true in R, one needs only to show that for each degree, its homogeneous pieces are contained in the left ideal of identities for that degree. This is a matrix theory problem which can be done by row operations. It is a finite problem since one has only a finite number of degrees to check.

The next question is how to compute the $L_1, L_2, \ldots, L_n, \ldots$. Let $P_1, P_2, P_3, \ldots$ be a sequence of linear homogeneous (non-associative) polynomials. Assume that the degrees are nondecreasing and that there are only a finite number of each degree. Let R be the free non-associative algebra defined by these identities. If $P_1, P_2, \ldots, P_r$ are all the identities of minimal degree, then the submodule of identities of minimal degree is found directly. At each stage, if L_i is known, L_i will have a finite set of generators which correspond to a finite set of (non-associative) polynomial identities of degree i. By pre- and post-multiplying these identities by an $i+1^{th}$ unknown, identities of degree i+1 are generated. By replacing each argument with the product of two elements, each polynomial also generates i more identities of degree i+1. Finally, from the original sequence of identities one includes all those of degree i+1. This finite set of identities generates L_{i+1}. This shows by induction how the L_i, i = 1,2,... can be constructed.

Theorem. Let R be a free nonassociative algebra over a field of characteristic 0 which is defined by a finite set of homogeneous polynomials. If P is any identity, it may be determined systematically if P holds in R.

It would be impossible to carry out the details of this proof for polynomials of large degree. This type of analysis is, however, useful as Regev showed in [5]. If one constructs the matrices of the representations of S_n, at n=9 they are as large as one could conveniently handle. S_9 has an irreducible representation of degree 216. S_{10} has an irreducible representation of degree 768. The numbers of ways to associate nonassociative products are the numbers of the Catalan sequence. The sequence begins 1,2,5,14,42,132,... . Products of degree greater than 5 are generally too big to work with.

As one expects, it will be impossible to perform these calculations for products of large degree. When working with small degrees, however, this system has great significance. Most nonassociative identities which have been studied are of degree 4 or less. Practically all identities proven are of degree 5 or less. Identities of higher degree are implied directly by identities of much lower degree. An example is Kleinfeld's $([a,b]^4,r,s) \equiv 0$ in alternative rings. A method which can handle identities of degree 9 or less is certainly a valuable tool. The number of associations of 9 elements is 1430; one shudders at the thought of calculating the module consisting of 1430 copies of the group ring of S_9. Instead, one uses intuition to pick functions to keep the number of summands manageable. In nonassociative work, expressions like [a,[b,(c,d,e)]], [a,(b,c,[d,e])], (a,b[c,[d,e]]), and (a,[b,c],[d,e]) are called saturated. They are identified by the property that if any element is replaced by the unit element of the ring, the expression is zero. In much nonassociative work, the given identities imply identities based on the saturated functions. One then needs only one copy of the group ring for each saturated function. This technique is used in classifying associator dependent algebras in [3] and simple (γ,δ) algebras in [2].

Another technique is to use identities modulo an additive subgroup. In (γ,δ) algebras, it is shown that the additive span of the associators is an ideal. This is done by showing that the functions a(b,c,d) and (a,b,c)d are identities modulo the span of the associators. This is done in [3].

In a Jordan algebra J, the span of the absolute zero divisors is an ideal of the algebra. One then defines an ascending chain of ideals $H_1 \subseteq H_2 \subseteq H_3 \ldots$ where H_{i+1}/H_i is spanned by all absolute zero divisors of J/H_i. If $H = \sum_{i=1}^{\infty} H_i$, then H is a nil ideal. If R is a right alternative ring, then R^+ (that is, R under "o", the symmetric product) is a Jordan algebra. The ideals H_i will be R^+ ideals, but not necessarily R ideals. H is useful because one can construct ideals of R inside H which are then nil. Initially, one constructs absolute zero divisors of R^+. Then, using those absolute zero divisors as identities, one constructs absolute zero divisors of R/H_1. This process continues. Each time one is working with identities modulo an additive subgroup of R. This technique was developed by Thedy in [6]. He did not use representation theory.

It is quite tedious to work with identities. It is hard to keep track of all the terms in a large identity. It is easier to work with matrices. Over the rational numbers, the matrices of the representations of S_n are integral. The row operations can be done on a computer.

The representation method does not require linear functions. As an exmaple, one can use $f(a,b,c) = (ab)(bc)(ca)+(bc)(ca)(ab)+(ca)(ab)(bc)$. For this function, I-(123) is an identity of f. Non-linear functions are used to process identities which involve products with repeated factors.

Identities of nonassociative algebras of arbitrary high degree can be processed by group representation. If $P_1, P_2, \ldots, P_n$ are polynomial identities over a field of characteristic 0, then it is possible to decide if a polynomial is a consequence of $P_1, P_2, \ldots, P_n$. If the degrees are ≤ 9, it is possible to actually compute results using computers, though one would usually have to reduce the problem in steps by using techniques such as saturated functions and identities modulo additive subgroups.

References

[1] I. Hentzel, "Processing identities by group representation," Monograph on Computers in Nonassociative Rings and Algebras, Amer. Math. Soc., January 1976.

[2] _______ and G. M. Piacentini, "Simple (γ,δ) algebras are associative," to appear J. Algebra.

[3] I. Hentzel, G. M. Piacentini, and D. Floyd, "Alternator and associator ideal algebras," to appear Trans. Amer. Math. Soc.

[4] N. Jacobson, "Structure of Rings," Amer. Math. Soc. Colloq. Publ., vol. 37, 1964.

[5] J. Olsson and A. Regev, "An application of representation theory to PI-algebras," to appear Proc. Amer. Math. Soc.

[6] A. Thedy, "Right alternative rings with Peirce decomposition," J. Algebra 37 (1975), 44-63.

RINGS WHOSE CYCLIC MODULES HAVE CERTAIN PROPERTIES AND THE DUALS

S. K. Jain

Ohio University
Athens, Ohio

The object of this article is to give a survey of the results for rings whose cyclic modules have certain properties, like, injectivity, quasi-injectivity, continuity, π-injectivity, rational completeness, projectivity and quasi-projectivity. The dual results for rings whose one-sided ideals are quasi-injective or quasi-projective are also given and open questions are stated. The contributors in this theory include Faith, Osofsky, Levy, Singh, Koehler, Mohamed, Ivanov, Hill, Courter, Klatt, Jain and others. We assume throughout that all rings have unity and all modules are right and unital.

In her dissertation Osofsky [36] proved the first important result in this direction for rings all of whose cyclic modules are injective.

<u>Theorem</u> (Osofsky [37]). If every cyclic R-module is injective then R is semisimple artinian.

Later she gave a simplified proof of the above theorem [38]. She claims that this result was also proved by Skornjakov [40] but that his proof had an error.

In his Ph.D. dissertation Cozzens [7] produced an example of a noetherian domain R in which every cyclic R-module is either free or semisimple. Following the example of Cozzens, Boyle ([3], [4]) considered rings in which proper cyclic modules are injective. She called these rings right PCI-rings and showed that noetherian right PCI-rings are either semisimple artinian or simple hereditary domains. Faith considered right PCI-rings without chain conditions and proved

Theorem (Faith [8]). A right PCI ring R is either semisimple, or a right semihereditary simple domain.

How close right PCI-domains come to being hereditary rings is an open question?

The example of Cozzens and the study of right PCI-rings led Goel, Jain and Singh to consider rings R for which every cyclic R-module is injective or projective. They proved the following theorem which, in particular, provides a simple proof of Faith's result that regular right PCI-ring is semisimple artinian.

Theorem (Goel-Jain-Singh [14]). If each cyclic R-module is injective or projective then $R = A \oplus B$ where A is semisimple artinian and B is a simple right semihereditary right Öre-domain whose every proper cyclic right module is injective.

Faith's interest in the study of right PCI-rings stems from the well-known cyclic decomposition of a module over a dedekind domain. One way to obtain the cyclic decomposition is to observe that for a dedekind domain R every proper cyclic module R/A is injective as R/A-module. Thus Faith [8] proposed to study the class of rings R with the condition:

(F) Every cyclic module $C \not\approx R$ is injective modulo its annihilator ideal.

But instead he studied right PCI-rings R which do satisfy the condition (F). Commutative rings with the condition (F) are precisely the pre-self-injective rings characterised by Klatt and Levy. They have shown

<u>Theorem</u> (Klatt-Levy [26]). A commutative ring R is pre-self-injective if and only if it is one of the following.

(1) An integral domain (necessarily a Prüfer domain) in which, for every maximal ideal M, R_M is an almost maximal rank 1 valuation domain; and every proper ideal is contained in only finitely many maximal ideals.

(2) The direct sum of a finite number of maximal rank 0 valuation rings. (Here R is also self-injective).

(3) An almost maximal rank 0 valuation ring.

(4) A local ring whose maximal ideal M has composition length 2 and satisfies $M^2 = 0$. (Here R is not self-injective.)

Right self-injective rings with the condition (F) have the property that each cyclic R-module is quasi-injective. These rings are studied by Ahsan [1] and are called qc-rings. Koehler has obtained a complete characterisation of qc-rings in the following theorem.

<u>Theorem</u> (Koehler [29]). For a ring R the following are equivalent:

(i) Each cyclic right R-module is quasi-injective.

(ii) $R = A \oplus B$ where A is semisimple artinian and B is a finite direct sum of rank 0 maximal valuation duo rings.

(iii) Each cyclic left R-module is quasi-injective.

Thus, the above theorem proves

<u>Theorem</u>. Right self-injective rings with the condition (F) are precisely qc-rings.

Jain, Singh and Symonds call a ring R a right PCQI-ring if each cyclic R-module $C \not\cong R$ is quasi-injective. Clearly rings with the

(F)-condition are right PCQI-rings. The following are their main results:

Theorem (Jain-Singh-Symonds [25]). A right PCQI-ring R is either prime or semiperfect. Further, if R is non-prime, non-local, then R is a right PCQI-ring iff either R is a qc-ring or R = $\begin{pmatrix} D & D \\ 0 & D \end{pmatrix}$ where D is a division ring.

Theorem Jain-Singh-Symonds [25]). A local right PCQI-ring with maximal ideal M is a right valuation ring or $M^2 = (0)$. In addition, if R is prime then R is a valuation domain.

Theorem (Jain-Singh-Symonds [25]). A right PCQI-domain is a right Öre-domain.

The above result was proved by Faith [8] for PCI-domains. If R is commutative then some of the results of Klatt and Levy [26] follow as a special case of the above results for PCQI-rings.

In an attempt to study rings R satisfying condition (F), Jain and Singh have obtained the following partial solution.

Theorem (Jain-Singh [23]). Let R be a ring satisfying condition (F). Then either R is a right Öre-domain or semiperfect. Further, a semiperfect ring R satisfies condition (F) iff R is a right PCQI-ring.

So the above theorem leaves the question of characterising right Öre-domains with condition (F) open.

By assuming the ring R to be duo, Koehler has shown the following:

Theorem (Koehler [30]). Let R be a duo domain. Then R satisfies condition (F) (equivalently, R is a right PCQI-ring, or R is a right pre-self-injective ring) iff

(1) Every proper ideal is contained in only finitely many maximal ideals, and

(2) For every maximal ideal M and every proper ideal A_M in the localization R_M, R_M/A_M is a maximal rank 0 duo valuation ring.

Analogous to continuous rings defined by Utumi [42], Mohamed and Bouhy [35] introduced the notion of a continuous module as a generalization of quasi-injective module. A module M is called continuous if it satisfies the following conditions:

(i) Every submodule of M is essential in some direct summand of M.

(ii) If a submodule A of M is isomorphic to a direct summand of M, then A itself is a direct summand of M.

A ring R is called a right cc-ring if every cyclic R-module is continuous. The following is a generalization of Koehler's theorem for qc-rings.

Theorem (Jain-Mohamed [21]). A semiperfect ring R is a right cc-ring iff $R = A \oplus B$ where A is semisimple artinian and B is a finite direct sum of right valuation right duo rings with nil radical.

Another generalization of Osofsky's theorem and Koehler's theorem is given by Goel and Jain. But first some terminology. Call an R-module M to be π-injective if for all R-modules A_1, A_2 of M with $A_1 \cap A_2 = (0)$, each projection $\pi_i : A_1 \oplus A_2 \to A_i$, $i = 1,2$ can be lifted to an endomorphism of M. Every quasi-injective module is π-injective but conversely π-injective modules need not be quasi-injective. For example, if R is a right Öre-domain which is not a division ring then R as a right R-module is π-injective but not quasi-injective. Some results on π-injective modules are:

<u>Theorem</u> (Goel-Jain [15]). (i) M is π-injective iff M is (V,R)-submodule of $\hat{M}$, the injective hull of M, where V is the subring of the ring of endomorphisms of $\hat{M}$ generated by all the idempotents.

(ii) Let R be a right noetherian ring and M be a π-injective R-module. Then M is a finite direct sum of indecomposable π-injective modules.

(iii) Let M,N be R-modules such that $M \times N$ is π-injective. Then M is N-injective and N is M-injective.

(iv) A prime ring R with $Z(R) = (0)$ is π-injective as an R-module iff R is self-injective or right Öre-domain.

The study of rings R over which each cyclic module is π-injective has been initiated by Goel and Jain [15]. The following theorems generalize the theorem of Osofsky [38] as well as that of Koehler [29]:

<u>Theorem</u> (Goel-Jain [15]). Let R be a right self-injective ring. Then each cyclic R-module is π-injective iff $R = A \oplus B$ where A is semisimple artinian and B is a finite direct sum of right self-injective right valuation rings.

<u>Theorem</u> (Goel-Jain [15]). Let R be a semiperfect ring. Then each cyclic R-module is π-injective iff $R = A \oplus B$ where A is semisimple artinian and B is a finite direct sum of right valuation rings.

<u>Theorem</u> (Goel-Jain [15]). Let R be a ring with right zero singular ideal such that each cyclic R-module is π-injective. Then $R = A \oplus B$ where A is semisimple artinian and B is a finite direct sum of right Öre-domains.

In this connection we ask: Is it true that a prime ring R all of whose cyclic R-modules are π-injective must have zero singular ideal? Since a right valuation ring, clearly, has the property that each cyclic module is π-injective, an example of a prime right valuation ring which is not a domain could settle this question in the negative.

Findlay and Lambek ([10], [11]) calls an R-module M a rational extension of N if N is a submodule of M and for each $x \in M$, $0 \neq y \in M$ there exists $r \in R$ such that $xr \in N$ and $yr \neq 0$. It is known that M is a rational extension of N iff for each submodule B of M which contains N, each $f \in \hom(B,M)$ such that $f(N) = 0$ is the zero mapping. A module M is rationally complete if M has no proper rational extensions. They have shown

Theorem (Findlay-Lambek [11]). M is rationally complete if and only if for each pair of R-modules A and B such that A is a submodule of B and that for each $b \in B$ and for each $0 \neq m \in M$, there exists $r \in R$ such that $br \in A$ and $mr \neq 0$, each $f \in \hom(A,M)$ can be extended to a $g \in \hom(B,M)$.

Thus, every injective module is rationally complete. Courter has given the following characterisation for a ring R whose R-modules are rationally complete.

Theorem (Courter [6]). Every R-module is rationally complete iff R is a finite direct sum of rings R_i isomorphic to the $n_i \times n_i$ matrix ring over a right and left perfect completely primary ring S_i.

Bland [2] has shown that every R-module is rationally complete if and only if every R-module is quasi-rationally complete in the following sense:

If N is a submodule of M such that (i) for each $x \in M$, $0 \neq y \in M$ there exists $r \in R$ such that $xr \in N$ and $yr \neq 0$ and (ii) M is contained in the quasi-injective hull of N, then M is called a quasi-rational extension of N. Further, a module M is called quasi-rationally complete if M has no proper quasi-rational extensions.

It is an open problem to study the class of rings R each of whose cyclic modules are rationally complete. Brown has proved some results for rings over which every simple module is rationally complete. Brown [5] has shown that the collection of rings over which every simple module is rationally complete contains all finite direct sums of matrix rings over duo rings.

Rings R for which a certain subclass of the cyclic R-modules, simple R-modules, are injective or projective have been of interest to various authors. For example, rings whose simple modules are injective, due to Villamayor, are called V-rings and have been studied extensively. We state below only two results. The reader is referred to Faith [9] for results on V-rings.

Theorem (Villamayor). For a ring R the following are equivalent:

(a) Each simple R-module is injective

(b) Each right ideal is the intersection of maximal right ideals

(c) For all modules M, the radical of $M = 0$.

Theorem (Cozzens [7]). Let k be a universal differential field with derivation D. The ring $R = k[y,D]$ has the following properties:

(a) R is a simple principal right and left ideal domain

(b) R is a V-ring.

(c) Up to isomorphism, R has a unique simple R-module.

Rings whose simple module are injective or projective have been studied by Zaks [44], Fuller [12] and others.

Theorem (Fuller [12]). Let R be a right artinian ring with Jacobson radical N. Then the following are equivalent:

(a) Every submodule of a quasi-projective R-module is quasi-projective.

(b) Every factor ring of R is hereditary.

(c) Every factor module of a quasi-injective R-module is quasi-injective.

(d) R is hereditary and $N^2 = 0$.

(e) Every simple module is injective or projective.

It is remarked by Fuller that the conditions (a), (b), (d) and (e) are equivalent over a perfect ring R. It is not known whether (a) and (c) are equivalent over any larger class of rings.

Rings over which every cyclic module is projective are trivially semisimple artinian. The study of rings for which every cyclic module is quasi-projective was initiated by Koehler. She calls a ring R a right q*-ring if each cyclic R-module is quasi-projective. Some of the main results are:

Theorem (Koehler [27]). (i) A semiperfect ring R is a right q*-ring iff every right ideal in the Jacobson radical of R is an ideal.

(ii) Let R_n be the $n \times n$ matrix ring over a ring R with $n > 1$. Then R_n is a right q*-ring iff R is semisimple artinian.

(iii) Let R be a right self-injective semiperfect right q*-ring. Then $R = A \oplus B$ where A is semisimple artinian and $B = \oplus \sum_{i=1}^{n} e_1 R$ is a finite direct sum of indecomposable right ideals $e_i R$ such that $e_i R \not\cong e_j R$ if $i \neq j$.

(iv) If R is a prime semiperfect q*-ring, then R is either simple artinian or R is local.

(v) Let R be a quasi-Frobenious ring. Then R is a right q*-ring iff R is a left q*-ring.

The study of rings with the dual property (rings in which every right ideal is quasi-injective) was begun by Jain, Mohamed and Singh. They call these rings with the above property right q-rings and have proved the following results.

Theorem (Jain-Mohamed-Singh [24]). (i) R is a right q-ring iff R is right self-injective and each essential right ideal is an ideal.

(ii) Let $n > 1$ be an integer. Then the $n \times n$ matrix ring R_n is a right q-ring iff R is semisimple artinian.

(iii) A prime ring R is a semisimple artinian iff R is a right q-ring.

(iv) A semi-prime ring R is a right q-ring iff $R = A \oplus B$ where A is semisimple artinian and B is a self-injective strongly regular ring.

Mohamed in his dissertation [32] studied various classes of q-rings. Some of his main results are:

Theorem (Mohamed [33]). (i) Let R be a semi local right q-ring. If eRe is not a division ring for any primitive idempotent $e \in R$, then R is a finite direct sum of local right self-injective duo rings.

(ii) Let R be a PF-ring (in the sense of Utumi [43]). If R is a q-ring then $R = A \oplus B$ where A is a quasi-Frobenious q-ring and B is a finite direct sum of local right self-injective duo rings with nonzero socle.

(iii) A right or left artinian right q-ring is also a left q-ring.

The next two theorems generalize Levy's [31] result for commutative noetherian rings whose homomorphic images are self-injective to the noncommutative case.

Theorem (Mohamed [34]). Let R be a nonprime right noetherian ring. Then every proper homomorphic image of R is a right q-ring if and only if

(a) $R = S \oplus T$ where S is semisimple artinian and T is a principal ideal duo ring with dcc, or

(b) If J denotes the Jacobson radical then R/J is artinian and J is a minimal ideal, or

(c) R is a local ring whose maximal right ideal M satisfies $M^2 = (0)$ and every proper homomorphic image of R contains at most one proper right (left) ideal.

Theorem (Mohamed [34]). Let R be a prime right noetherian ring with the property that every homomorphic image is a right q-ring. Then,

(i) Every ideal of R is a product of prime ideals, and

(ii) For every nonzero prime ideal of R, R/P is a division ring.

Hill [16] in his dissertation, has, among other results, characterized semiperfect q-rings.

Theorem (Hill [17]). Let R be semiperfect. If R is a right q-ring then R is a direct sum of a semisimple artinian ring and a right self-injective basic ring. Further, a right self-injective basic ring which is semiperfect need not be right q-ring.

Theorem (Hill [17]). Let R be an injective cogenerator as a right R-module. Then R is a right q-ring iff R is a finite direct sum of indecomposable rings R_i of the following three types.

(a) R_i is simple artinian

(b) R_i is local, and for each $x \in R_i$, $xR_i = R_i x$.

(c) R_i is basic, not local, and for each pair e,f of orthogonal idempotents $eR_i f \subseteq$ right socle of R_i = left socle of R_i.

Corollary (Hill [17]). A right artinian ring is a right q-ring iff it is a finite direct sum of indecomposable rings R_i of the above three types (a), (b) or (c).

In his dissertation Ivanov [18] has determined the structure of indecomposable non-local q-rings. He has shown that there are only two types (both artinian) of indecomposable non-local q-rings and has represented them as matrix rings, one of which is denoted by H(m,D,V) and defined as follows: For any integer $m \geq 2$, any division ring D, and any null D-algebra V whose left and right D-dimensions are both equal to one, H(m,D,V) denotes the ring of all $m \times m$ matrices whose only non-zero entries are arbitrary elements of D along the diagonal, and arbitrary elements of V at the places (2,1), ..., (m,m-1), (1,m).

Theorem (Ivanov [18]). Let R be an indecomposable non-local right q-ring. Then R is artinian and is either simple artinian or is of the form H(m,D,V). Conversely, every simple artinian ring or every ring of the form H(m,D,V) is a q-ring.

Conjecture (Ivanov [19]). Every right q-ring is a direct sum of a finite number of indecomposable non-local right q-rings and a right q-ring all of whose idempotents are central.

The conjecture is true iff every right ideal in a q-ring is contained in an ideal which is generated by a central idempotent and is minimal with respect to those properties iff the intersection of all right ideals which have (in the ring) a common homomorphic image disjoint from them is nonzero (c.f. [17]).

Some of the relationships between q-rings, q*-rings and qc-rings are given by Koehler.

Theorem (Koehler [29]). If R is a right qc-ring then R is a right q-ring and a right q*-ring.

Theorem (Koehler [27]). (i) If R is a semiperfect right q-ring then R is a left q*-ring.

(ii) If R is a left cogenerator and a left q-ring then R is a left q*-ring.

The next theorem proved by Koehler reproves a theorem of Mohamed that a right artinian right q-ring is a left q-ring.

Theorem (Koehler [27]). Let R be a quasi-Frobenious ring. Then the following statements are equivalent.

(i) R is a left q-ring.

(ii) R is a right q-ring.

(iii) R is a left q*-ring.

(iv) R is a right q*-ring.

We know that rings for which each right ideal is projective are hereditary rings, and left perfect rings are those for which each module has a projective cover. A class of commutative rings in which each ideal has a projective cover has been considered by Snider in the following theorem.

<u>Theorem</u> (Snider [41]). Let R be a commutative indecomposable ring. Then each ideal of R has a projective cover iff R is either perfect, dedekind or local noetherian.

It remains open to study other classes of rings with the property that each right ideal has a projective cover.

Recall R is called a right qc-ring if each cyclic R-module is quasi-injective. The class of rings with the dual property that each right ideal is quasi-projective has been considered by Jain and Singh. They have called a ring R a right qp-ring if each of its right ideals is quasi-projective. Clearly, all right hereditary rings are right qp-rings. However, the class of commutative principal ideal artinian rings which are not direct sum of fields distinguishes qp-rings from hereditary rings. The following results giving the structure of right and left perfect right qp-rings R are obtained by Jain-Singh. N will denote the Jacobson radical.

<u>Theorem</u> (Jain-Singh [22]). Let R be a local ring. Then R is a right qp-ring iff $N^2 = (0)$ or R is a principal right ideal ring with dcc such that for each $a \in R$, ann a is an ideal in R.

<u>Theorem</u> (Jain-Singh [22]). If R is a perfect right qp-ring and T is the sum of all those indecomposable right ideals of R which are not projective then T is an ideal of R and $N = T \oplus L$; L is a right ideal of R such that every right subideal of L is projective, R/T is hereditary, and R is hereditary iff $T = (0)$.

<u>Theorem</u> (Jain-Singh [22]). If R is a perfect right qp-ring then $R = \begin{pmatrix} S & M \\ 0 & T \end{pmatrix}$ where S is hereditary, T is a right qp-ring which is a finite direct sum of local rings and M is (S,T)-bimodule such that

${}_SM$ is projective. In particular, a perfect right qp-ring is semi-primary.

Question: Is M_T quasi-projective?

Theorem (Jain-Singh [22]). (i) Let R be an indecomposable ring such that it admits a faithful projective injective right module. Then R is a right qp-ring iff R is a local principal right ideal ring such that for each $a \in R$, ann a is an ideal or R is a right hereditary ring with dcc.

(ii) Let R be an indecomposable quasi-Frobenious ring. Then R is a right qp-ring iff each homomorphic image of R is a q-ring.

Theorem (Jain-Singh [22]). If a right ideal A of a right qp-ring is not projective then the projective dimension of A is infinite. Thus gl. dim. R = 0, 1, or ∞.

It has been shown by Goel-Jain [13] and Singh-Mohammad [39] independently that all the above results are true for right perfect right qp-rings. The study of right qp-rings other than for perfect rings remains open. Indeed this leads to another important problem, namely, what is the structure of quasi-projective modules over some other class of rings? In particular, quasi-projective modules over local rings have not been studied. Finitely generated quasi-projective modules over semiperfect rings have been characterized by Koehler [28].

References

[1] J. Ahsan, "Rings all of whose cyclic modules are quasi-injective," Proc. London Math. Soc. 27 (1973), 425-439.

[2] Paul E. Bland, "Rings over which every module is quasi-rationally complete," J. of Natural Sciences and Math. 10 (1970), 269-274.

[3] Ann K. Boyle, "Hereditary QI-rings," Trans. Amer. Math. Soc. 192 (1974), 115-120.

[4] ________, Ph. D. thesis, Rutgers, The State University, New Brunswick, New Jersey, 1971.

[5] S. H. Brown, "Rings over which every simple module is rationally complete," Can. J. Math. 25 (1973), 693-701.

[6] R. Courter, "Finite direct sums of complete matrix rings over perfect completely primary rings," Can. J. Math. 21 (1964), 430-446.

[7] J. Cozzens, Ph. D. thesis, Rutgers, The State University, New Brunswick, New Jersey, 1969.

[8] C. Faith, "When are proper cyclic injective," Pacific J. Math. 45 (1973), 97-112.

[9] ________, "Algebra: Rings, Modules, and Categories I," Springer-Verlag, New York (1973).

[10] G. Findlay and J. Lambek, "A generalized ring of quotients I," Can. Math. Bull. 1 (1958), 77-85.

[11] ________, "A generalized ring of quotients II," Can. Math. Bull. 1 (1958), 155-167.

[12] K. Fuller, "Relative projectivity and injectivity classes determined by simple modules," J. London Math. Soc. (2), 5 (1972), 423-431.

[13] S. C. Goel and S. K. Jain, "Semiperfect rings with quasi-projective left ideals," Math. J. Okayama Univ., to appear.

[14] ________ and S. Singh, "Rings whose cyclic modules are injective or projective," Proc. Amer. Math. Soc. 53 (1975), 16-18.

[15] V. K. Goel and S. K. Jain, "π-injective modules and rings whose cyclics are π-injective," Preprint, 1976.

[16] D. A. Hill, Ph. D. thesis, University of Oregon, Eugene, Oregon, 1970.

[17] _______, "Semi-perfect q-rings," Math. Ann. 200 (1973), 113-121.

[18] G. Ivanov, Ph. D. thesis, Australian National University, Canberra, 1972.

[19] _______, "Non-local rings whose ideals are quasi-injective," Bull. Austral. Math. Soc. 6 (1972), 45-52.

[20] _______, "Addendum to 'Non-local rings whose ideals are quasi-injective,'" Bull. Austral. Math. Soc. 12 (1975), 159-160.

[21] S. K. Jain and Saad Mohamed, "Rings whose cyclics are continuous," preprint, 1976.

[22] S. K. Jain and S. Singh, "Rings with quasi-projective left ideals," Pacific J. Math. 60 (1975), 169-181.

[23] _______, "A generalization of PCI-rings," preprint, 1976.

[24] S. K. Jain, S. H. Mohamed and S. Singh, "Rings in which every right ideal is quasi-injective," Pacific J. Math. 31 (1969), 73-79.

[25] S. K. Jain, S. Singh and R. G. Symonds, "Rings whose proper cyclic modules are quasi-injective," to appear in Pacific J. Math.

[26] G. Klatt and L. Levy, "Pre-self injective rings," Trans. Amer. Math. Soc. 137 (1969), 407-419.

[27] A. Koehler, "Rings for which every cyclic module is quasi-projective," Math. Ann. 189 (1970), 311-316.

[28] _______, "Quasi-projective and quasi-injective modules," Pacific J. Math. 36 (1971), 713-720.

[29] _______, "Rings with quasi-injective cyclic modules," Quart. J. Math. Oxford 25 (1974), 51-55.

[30] _______, "Pre-self-injective duo rings," Preprint, 1975.

[31] L. Levy, "Commutative rings whose homomorphic images are self-injective," Pacific J. Math. 18 (1966), 149-153.

[32] S. Mohamed, Ph. D. thesis, University of Delhi, India, 1969.

[33] _______, "q-rings with chain conditions," J. London Math. Soc. 2 (1970), 455-460.

[34] _______, "Rings whose homomorphic images are q-rings," Pacific J. Math. 35 (1970), 727-735.

[35] _______ and T. Bouhy, "Continuous modules," Notices Amer. Math. Soc. 23 (1976), A-478.

[36] B. Osofsky, Ph.D. thesis, Rutgers, The State University, New Brunswick, New Jersey, 1964.

[37] _______, "Rings all of whose finitely generated modules are injective," Pacific J. Math. 14 (1964), 645-650.

[38] _______, "Noninjective cyclic modules," Proc. Amer. Math. Soc. 19 (1968), 1383-1384.

[39] S. Singh and A. Mohammad, "Rings in which every finitely generated left ideal is quasi-projective," to appear in Math. Student.

[40] L. A. Skornjakov, "Rings with injective cyclic modules," Soviet Mathematics 4 (1963), 36-38.

[41] R. L. Snider, "Rings whose ideals have projective covers," to appear in Archiv. der Math.

[42] Y. Utumi, "On continuous rings and self-injective rings," Trans. Amer. Math. Soc. 118 (1965), 158-173.

[43] _______, "Self-injective rings," J. Algebra 6 (1967), 167-173.

[44] A. Zaks, "Simple modules and hereditary rings," Pacific J. Math. 26 (1968), 627-630.

SURVEY OF RINGS WHOSE SUBGROUPS ARE SUBRINGS OR IDEALS

John D. O'Neill

University of Detroit
Detroit, Michigan

In [6] L. Redei described those associative rings whose additive subgroups are ideals. Here we give a complete description of associative rings whose additive subgroups are subrings. A ring with this property, called S, is designated a S-ring. For example, the ring of integers Z is a S-ring while the ring of rational numbers Q is not. All rings herein are associative, all groups are abelian, and all constants are integers. A ring R will be primary, torsion, torsion-free, mixed, etc. exactly if its additive group R^+ is primary, etc. By the rank of R, r(R), we mean the rank of R^+. Other terminology is essentially that of [4]. Proofs are omitted and may be found in [5]. However, a ring R is a S-ring if and only if, for every x and y in R, xy is a linear combination of x and y. This is the key to all proofs.

1. Torsion Rings

Proposition 1.1. If R is a p-primary S-ring, then R is bounded or R is a zero-ring.

Proposition 1.2. Let R be a ring on $G = \oplus\langle x_i\rangle$, a bounded p-group. Then R has property S iff there are constants u_i and v_i for each i such that:

(1) $x_i x_j = v_j x_i + u_i x_j$, for every i and j; or if $2^n G = 0$ and $r(2^{n-1}G) = 2$ for some n then

(2) $x_i x_j = v_j x_i + u_i x_j$, for $i \neq j$, and $x_i^2 = (u_i + v_i + 2^{n-1})x_i$, for every i.

How can a S-ring be constructed on a bounded primary group? In general, if group $G = \oplus\langle x_i\rangle$, a ring (not necessarily associative) can be constructed on G in the following way (see Theorem 120.1 of [4]). For each i and j let $x_i x_j$ be an element in G subject to the sole condition: $o(x_i x_j) \leq \min(o(x_i), o(x_j))$ and define multiplication on the rest of G linearly. These conditions are also necessary. We now have the following theorem.

Theorem 1.3. Let $G = \oplus\langle x_i\rangle$ be a bounded p-group where $o(x_i) = n_i$. A S-ring can be constructed on G by setting

(1) $x_i x_j = v_j x_i + u_i x_j$ and

(2) $(\Sigma a_i x_i)(\Sigma b_j x_j) = \sum_{i,j} a_i b_j x_i x_j$, where

(3) $n_i u_i G = n_i v_i G = 0$ and

(4) $u_i v_j x_k = 0$ unless $i = j = k$.

If $p \neq 2$, any p-primary S-ring, not a zero-ring, is of this type.

A S-ring on a 2-group, however, need not satisfy condition 1 of the preceding theorem. An example is the ring-direct sum $Z/(2) \oplus Z/(2)$. The following theorem will allow for this ring.

Theorem 1.4. Let G be as in (1.3) where $2^n G = 0$ and $r(2^{n-1}G) = 2$ for some n. A S-ring R can be constructed on G by setting

(1') $x_i x_j = v_j x_i + u_i x_j$ for $i \neq j$

$x_i^2 = (u_i + v_i + 2^{n-1})x_i$ for every i,

where (5) all u_i and v_i are even,

and satisfying conditions 2, 3, and 4 of (1.3). A S-ring on a 2-group, not a zero-ring, has this construction or that of (1.3).

At first glance, the construction of primary S-rings might seem unduly complicated, but this is not so. Suppose G is the group given in (1.3) with least bound p^N. We can satisfy the conditions of the theorem by, for each i, letting $u_i = r_i$, $v_i = 0$ where p^N divides $r_i n_i$. Another example shows that all u_i and v_i can be non-zero. Let R be the ring on the group $\langle x\rangle \oplus \langle y\rangle$ where $o(x) = 81 = o(y)$, $x^2 = 27x$, $y^2 = 27y$, $xy = 9x + 18y$, and $yx = 18x + 9y$. This ring also has property S.

Note. In (1.3) we may assume that each x_i is nilpotent except perhaps one of them which is then idempotent. If J is the radical of the ring R, then J is a nil ideal and R = J or R/J is isomorphic to Z/(p). In (1.4) either R = J a nil ideal or else R is isomorphic to the ring-direct sum of two copies of Z/(2).

We turn now to torsion S-rings in general.

Theorem 1.5. Let G be a torsion group with primary decomposition, $G = \bigoplus_p G_p$. A ring R on G has property S if and only if, for each p, the subring R_p on G_p has property S.

2. Torsion-free Rings

Theorem 2.1. Let G be a torsion-free group of the form $H \oplus \langle z\rangle$. A S-ring can be constructed on G by choosing u and s such that $u(s-u) = 0$ and, for $x,y \in H$, setting $(x + az)(y + bz) = bux + a(s-u)y + absz$. Any torsion free S-ring, not a zero ring, has such a structure.

3. Mixed Rings

Let R be a mixed ring and T its torsion subring. It is well-known that T is an ideal. If R is a S-ring, it also happens that $T^2 = 0$. If S-ring R is not a zero-ring, then R/T may or may not be a zero-ring. In the latter case R^+ has the form $H \oplus \langle z \rangle$ where $H \supseteq T$ and $H^2 = 0$. The next theorem describes all S-rings on such groups.

Theorem 3.1. Let $G = H \oplus \langle z \rangle$ be a mixed group where $H \supseteq T$. A S-ring R can be constructed on G by defining:

(1) if $H \neq T$, $(x + az)(y + bz) = a(s-u)y + bux + absz$ for $x,y \in H$ where $u(s-u) = 0$;

(2) if $H = \oplus T_i$, $(\Sigma x_i + az)(\Sigma y_i + bz) = a\Sigma(s-u_i)y_i + b\Sigma u_i x_i + absz$ for $x_i, y_i \in T_i$ where $\oplus T_i$ is a primary decomposition of T and $u_i(s-u_i)T_i = 0$ for each i.

If R is a mixed S-ring and R/T is not a zero-ring, then R has the above structure with $s \neq 0$.

If R is a S-ring, it may happen that R/T is a zero-ring but R is not. In this case the additive group of R/T has rank one but may not be cyclic. Hence, a few words on mixed groups of torsion-free rank one are in order. Let G be such a group and $T = \oplus T_i$, $i \in I$, its torsion subgroup (T_i is the p_i-component of T where p_i is a prime number). Let z be a torsion-free element in G. We partition I into subsets J and K such that i is in J or K exactly if the p_i-height of z + T in G/T is infinite or finite, respectively. The latter height is designated n_i. In this case G has a generating set:

$A = \{T_i, z_{jn}, z_k\}$, $i \in I$, $j \in J$, $k \in K$, $n = 1,2,\ldots$ where

$$z = p_j^n z_{jn} + t_{jn},\ t_{jn} \in T_j$$

$$(*) \qquad = p_k^{n_k} z_k + t_k,\ t_k \in T_k.$$

Here $z_k = z$ if $n_k = 0$. We write $G = (A,*)$.

We may now define a S-ring on the above-mentioned group.

Theorem 3.2. Let $G = (A,*)$. A S-ring R can be constructed on G in the following manner. For each i in I, choose u_i such that $u_i^2 T_i = 0$ and, if $i \in J$, $u_i = 0$. For each i and $k \neq i \neq j$, choose u_{ki} and u_{ji} such that $u_{ki} = 0 = u_{ji}$ if T_i is unbounded and $p_k u_{ki} \equiv 1 \equiv p_j u_{ji}$ (mod $p_i^{N_i}$) if T_i has least bound $p_i^{N_i}$. Multiplication of G is defined by the following rules:

(1) $T^2 = 0$ and $xy = -yx$ for $x,y \in A$.

(2) For $x_i \in T_i$, z_k, and z_{jn}, $x_i z_k - u_{ki}^{n_k} p_i^{n_i} u_i x_i$ and

$x_i z_{jn} = u_{ji}^n p_i^{n_i} u_i x_i$ $(n_i = 0$ if $i \in J)$.

(3) For $i \in K$ and z_k, z_{jn}, $z_i z_k = u_{ik}^{n_i} u_k t_k - u_{ki}^{n_k} u_i t_i$ and

$z_i z_{jn} = - u_{ji}^n u_i t_i$.

(4) For $i \in J$ and z_{im}, z_{jn}, $z_{im} z_{jn} = 0$.

(5) Multiplication on G is linear relative to A.

If R is a S-ring and R/T but not R is a zero-ring, then R has the above structure.

4. Special Properties

In this section we wish to describe some particular S-rings. Cummutative S-rings are easily decided by examining the theorems of the previous sections. We do not list them here. We do state the following theorems.

<u>Theorem 4.1</u>. A S-ring R has an identity if and only if R is isomorphic to Z, to Z/(2) ⊕ Z/(2) (ring-direct sum), or to ⊕ $Z/(p^{n_p})$ summed over a finite number of distinct primes p where n_p depends on p.

<u>Theorem 4.2</u>. The rings Z and Z/(p) are the only integral domains with property S.

<u>Theorem 4.3</u>. A non-trivial S-ring R is regular if and only if it has the form ⊕ Z/(p) (ring-direct sum) where each p ≠ 2 (p=2) occurs at most once (twice).

<u>Theorem 4.4</u>. A S-ring R is Artinian if and only if it is a torsion ring of finite rank.

<u>Theorem 4.5</u>. A S-ring R is Noetherian if and only if R^+ is a direct sum of a finite number of cyclic groups.

We also characterize those rings whose additive subgroups are ideals. The characterization is not new (see [6]) but is easily deduced from earlier theorems of this paper. In the following T_i is the p_i-component of T for prime number p_i, ⊕ signifies ring-direct sum, and (z), (z_i) signify the subring generated by z, z_i, respectively.

<u>Theorem 4.6</u>. A ring R has the property that every additive subgroup is an ideal if and only if

(1) R is a zero-ring,

(2) $R = T \oplus (z)$ and $T^2 = 0 = sT$ where $z^2 = sz$, or

(3) $R = \oplus T_i$ and, for $T_i^2 \neq 0$, $T_i = H_i \oplus (z_i)$ and $H_i^2 = 0 = s_i H_i$ where $z_i^2 = s_i z_i$.

References

[1] R. A. Beaumont, "Rings with additive group which is the direct sum of cyclic groups," Duke Math. J., 15 (1948), 367-369.

[2] _______ and R. S. Pierce, "Torsion-free rings," Illinois J. Math., 5 (1961), 61-98.

[3] L. Fuchs, "Abelian Groups," Publ. House Hungar. Acad. Science, Budapest, 1958.

[4] _______, "Infinite Abelian Groups," Academic Press, N.Y., Vol. I (1970), Vol. II (1973).

[5] J. D. O'Neill, "Rings whose additive subgroups are subrings," Pacific J. Math. (to appear).

[6] L. Redei, "Die Vollidealringe," Monatshefte fur Mathematik, 56 (1952), 89-95.

CLASSICAL LOCALIZATION AT PRIME IDEALS OF FULLY BOUNDED NOETHERIAN RINGS

F. L. Sandomierski

Kent State University
Kent, Ohio

Introduction

In [67, Goldie] initiated localization at a prime ideal of two-sided not necessarily commutative Noetherian ring and gave sufficient conditions for the localization to be classical.

If S is a semiprime ideal of a right Noetherian ring R and C(S) is the set of all elements of R whose images are regular elements in R/S, then Lambek-Michler [73] have shown that when S is a prime ideal, C(S) and E(R/S), the R-injective hull of the right R-module R/S, determine the same torsion theory. Subsequently, Jategaonkar [73], Lambek-Michler [74] have shown that the aforementioned torsion theories coincide when S is a semiprime ideal. In Jategaonkar [74] and Lambek-Michler [74] necessary and sufficient conditions for C(S) to be a right Ore set were given. In Cozzens-Sandomierski [76], henceforth C - S [76] necessary and sufficient conditions for C(S) to be a right Ore set were given, many of which were internal to the ring R and some module theoretic.

It is the object of this paper to sharpen the results of C - S [76] for a fully bounded Noetherian ring.

1. Preliminaries and Notation

Throughout this presentation R will be a right Noetherian ring and S will denote a semiprime ideal of R. The R-injective hull of a right R-module M_R will be denoted by E(M) and E = E(R/S) and Λ is the R-endomorphism ring of E_R.

Associated with S is the multiplicative set

$$C(S) = \{c \in R \mid x\,c \in S \text{ or } c\,x \in S \Rightarrow x \in S\}$$
$$= \{c \in R \mid c + S \text{ is regular in } R/S\}$$

and the idempotent topologizing filter

$$\Phi_S = \{I_R \subseteq R_R \mid a^{-1}\, I \cap C(S) \neq \emptyset \quad a \in R\}$$

where

$$a^{-1}I = \{r \in R \mid a\, r \in I\}.$$

For K a right ideal of R, the right closure of K with respect to S is $\rho_S(K) = \{x \in R \mid x\, I \subseteq K \text{ for some } I \in \Phi_S\}$. Lambek-Michler [74] and Jategaonkar [73] have shown that $I \in \Phi_S$ if and only if $\mathrm{Hom}_R(R/I,E) = 0$. $\mathcal{T}_S$, $\mathcal{F}_S$ will denote the Φ_S-torsion and Φ_S-torsion free classes respectively of the aforementioned torsion theory.

If X is a module over a ring T and $Y \subseteq X$, then $\mathrm{Ann}_T\, Y$ is the set of all elements of T which annihilate each element of Y. If $A \subseteq T$, then $\mathrm{Ann}_X\, A$ is the set of all elements of X which are annihilated by each element of A.

Since $\mathrm{Ann}_E\, S$ is an R/S injective hull of the right R/S module R/S, it is isomorphic to the classical ring of quotients of R/S as a right R/S or R module, see Lambek [66]. The classical ring of right quotients of R/S will be denoted by Q and will be identified with $\mathrm{Ann}_E(S)$ in what follows.

The following results for a semiprime ideal of R may be found in Müller [76].

(i) S is an irredundant intersection of prime ideals $\{P_1, P_2, \ldots, P_n\}$ each of which is a minimal prime ideal over S, and $Q = Q_1 + Q_2 + \ldots + Q_n$ is a ring decomposition into simple Artinian ring Q_i with $\mathrm{Ann}_R(Q_i) = P_i$, and Q_i is the classical ring of right quotients of R/P_i, $i = 1,2,\ldots,n$.

(ii) $\Phi_S = \bigcap_{i=1}^{n} \Phi_{P_i}$

(iii) Λ is a semiperfect ring and $J(\Lambda)$, the Jacobson radical of Λ, consists of all endomorphisms with essential kernel.

(iv) $C(S)$ is a right Ore set in R if and only if $\mathrm{Ann}_E(c) = 0$ for every $c \in C(S)$.

S is said to be right localizable when $C(S)$ is a right Ore set in R. In this case R has a ring of right fractions with respect to $C(S)$ and will be denoted by R_S. If Φ is an idempotent topologizing filter of right ideals of R, then a right ideal C of R is Φ-closed if R/C is Φ-torsion free. C is Φ-critical if C_R is maximal among Φ-closed ideals and $C \neq R$, see Lambek-Michler [73].

For the torsion theory associated with S we establish the following notation. For a right R-module M_R $t_S(M)$ will denote the Φ_S-torsion submodule of M. The class of Φ_S-torsion respectively Φ_S-torsion free modules will be denoted by $\mathcal{T}_S$ respectively $\mathcal{F}_S$.

For a module M_R, M^S denotes the S-primary component of M, which is the set of all elements of M which are annihilated by some power of S. Also M_n will denote the set of elements of M annihilated by S^n for $n = 1,2,\ldots$. The Krull dimension of M_R will be denoted $\kappa(M_R)$. The results on Krull dimension in Jategaonkar [74] will be utilized freely in this presentation.

2. Artinian Rings

In this section we give some characterizations of right localizable semiprime ideals of a right Artinian ring and characterize those left and right Artinian rings all of whose maximal ideals are right localizable.

Theorem 2.1. Let R be a right Artinian ring and S a semiprime ideal with index of idempotency n (i.e. $S^n = S^{n+1} = \dots$.) then the following statements are equivalent.

1. S is right localizable
2. If N is a maximal ideal of R not containing S, then $N \cap S = NS$
3. S satisfies the right AR property
4. $S^n = Re$ for some idempotent $e \in R$.

Proof. 1. implies 2. Since R is right Artinian prime ideals are maximal, so $S = M_1 \cap M_2 \cap \dots \cap M_r$, M_i maximal ideals. Now as S is not contained in N, $N \neq M_i$ for all i hence $N \in \Phi_{M_i}$ for all i as is easily seen and therefore $N \in \bigcap_{i=1}^{r} \Phi_{M_i} = \Phi_S$. By C - S [76, Theorem 2.3] $N \cap S \subseteq \rho_S(NS) = NS$, where the proof of the last equality can be found in the proof of Corollary 1 C - S [67, Theorem 3.7]. The inclusion $NS \subseteq N \cap S$ is trivial so 2. follows.

2. implies 1. In view of C - S [76, Theorem 3.4] it is sufficient to show that $E/Q \in F_S$. Now $\mathrm{Tor}_1^R(R/N, R/S) \simeq N \cap S/NS$ for any ideal N of R, thus $\mathrm{Tor}_1^R(R/N, R/S) = 0$ for each maximal ideal N not containing S and therefore $\mathrm{Tor}_1^R(U, R/S) = 0$ for each simple right R-module U not isomorphic to a submodule of R/S. Now by the associativity formula Cartan-Eilenberg [56] it follows that $\mathrm{Ext}_R^1(V, R/S) = 0$ for each simple right R-module V not isomorphic to submodule of R/S. Now if

$E/Q \notin F_S$, then since R is right Artinian the torsion submodule of E/Q would contain a simple module V_R which is necessarily not isomorphic to a submodule of R/S. Since Q = R/S, this would give to an exact sequence

$$0 \to R/S \to X \to V \to 0$$

of right R-modules. Since this sequence is not split this contradicts that $\mathrm{Ext}_R^1(V, R/S) = 0$, and 1. follows. The equivalence of 1. and 3. may be found in C - S [76, Theorem 3.7, Corollary 3].

3. implies 4. Since S has the right AR property E is S-primary by C - S [76, Theorem 3.8] and clearly E is an injective cogenerator over the ring $\overline{R} = R/S^n$, and $E_R \simeq \mathrm{Hom}({}_R\overline{R}_{\overline{R}}, E_{\overline{R}})$. So by Morita [70, lemma 1.3] it follows that ${}_R\overline{R}$ is flat as a left R-module and therefore projective by Bass [60] and 4. follows.

4. implies 3. Since $\mathrm{Hom}({}_R\overline{R}_{\overline{R}}, E_{\overline{R}}) \simeq E_R^S$ and ${}_R\overline{R}$ is a R-flat again by Morita [70]. E_R^S is injective so $E^S = E$ and by C - S [76, Theorem 3.8] and the fact that $IS = \rho_S(IS)$ for each right ideal of R, S has the right AR property and 3. follows and the theorem is proved.

Theorem 2.2. For a left and right Artinian ring R, the following statements are equivalent.

1. Every maximal ideal is right localizable.
2. $R = R_1 + R_2 + \ldots + R_n$ a ring direct sum with each R_i a full matrix ring over a local Artinian ring Λ_i, $i = 1,2,\ldots,n$.
3. Every maximal ideal is left localizable.

Proof. By Theorem 2.1 statement 1. is equivalent to $N \cap M = NM$ for distinct maximal ideals M, N. Since this condition is symmetric the equivalence 1. and 3. follows.

1. implies 3. Let $M_1, M_2, \ldots, M_n$ be the distinct maximal ideals of R, then $J = M_1 \cap M_2 \cap \ldots \cap M_n$ is the Jacobson radical of R. Let m be

an integer greater than the index of nilpotence of J and greater than the index of idempotence of each M_i. Now using the fact that each M_i has the right AR property and C - S [76, Theorem 3.8] it follows that $A = M_1^m \cap M_2^m \cap \ldots \cap M_n^m = M_1^m M_2^m \ldots M_n^m$. As was earlier observed that 1. is equivalent to maximal ideals commuting we have that $A = (M_1 M_2 \ldots M_n)^m \subseteq J^m = 0$. Since the ideals $M_1^m, \ldots, M_n^m$ are clearly comaximal in pairs and their intersection is zero the Chinese Remainder Theorem yields a ring decomposition

$$R = R_1 + R_2 + \ldots + R_n$$

with $R_i \simeq R/M_i^m$ as rings.

Now R_i is left and right Artinian ring with a unique simple right R-module up to isomorphism. Now if $T = e_1T + \ldots + e_kT$ is a decomposition of a right Artinian ring T with e_j primitive idempotents, each e_jT is the projective cover of a simple right T-module [60]. Thus, if T has a unique simple right T-module up to isomorphism all of the e_jT are isomorphic [60]. Thus T is ring isomorphic to $\mathrm{Hom}(T_T, T_T)$ which is isomorphic to the full $k \times k$ matrix ring over $\mathrm{Hom}_T(e_1T, e_1T)$ which is a local Artinian ring and the implication 1. implies 2. follows. The implication 2. implies 1. is straightforward and is left to the reader.

3. Fully Bounded Noetherian Rings

In this section we sharpen the results of C - S [76] for fully bounded Noetherian rings. The results of Jategaonkar [74] will play a central fole here. The following result is due to Jategaonkar [74, Lemma 5.2]. It is stated here and a proof given which is different.

<u>Theorem 3.1</u> [74, Jategaonkar]. If S is a semiprime ideal of an FBN ring so that the Krull dimension of $R/P = \alpha$ for every prime ideal P

associated with S, then S/S^n is right localizable in R/S^n for $n = 1,2,\ldots$.

Proof. By C - S [76, Theorem 2.4] it is sufficient to show that S/S^2 is right localizable in R/S^2. Thus, we may assume that $S^2 = 0$. Let $S = P_1 \cap P_2 \cap \ldots \cap P_n$ be the irredundant intersection of the associated prime ideals of S, then $E = E_1 \oplus E_2 \oplus \ldots \oplus E_n$ where $E = E(R/S)$ and $E_i = E(R/P_i)$ for $i = 1,2,\ldots,n$. Denote $Q_i = \mathrm{Ann}_E(P_i) \subseteq E_i$. Since $E/Q \simeq E_1/Q_1 \oplus \ldots \oplus E_n/Q_n$, then $t_S(E/Q)$, the Φ_S-torsion submodule of E/Q is isomorphic to the direct sum of $t_S(E_i/Q_i)$ the Φ_S-torsion submodules of E_i/Q_i. Now let P be any P_i, i-fixed and $t_S(E_i/Q_i) = X_i/Q_i$. Let B be an ideal of R that is maximal with respect to the property that there exists $x \in X_i - Q_i$ with $x B \subseteq Q_i$. It is clear that B is a prime ideal of R.

Claim 1. $P_j \subsetneq B$ for some $1 \leq j \leq n$.

Since $S^2 = 0$, $S \subseteq B$ so $P_1P_2 \ldots P_n \subseteq B$ and as B is prime some $P_j \subseteq B$. Now if $P_j = B$, then $\bar{x} = x + Q_i$ generates a faithful R/P_j submodule of X_i/Q_i. Since $\bar{x}R$ is Φ_S-torsion, $\bar{x}R$ is Φ_{P_j}-torsion an impossibility by Jategaonkar [74].

Claim 2. $B \in \Phi_S$.

Since $P_j \subsetneq B$, $B \not\subseteq P_k$ for all k since $P_1,P_2,\ldots,P_n$ are incomparable.

Claim 3. $x(B \cap P) \neq 0$.

If $0 = x(B \cap P)$, then $xPB = 0$. Since $B \in \Phi_S$, $xP = 0$, so $x \in Q_i$ an impossibility.

Claim 4. If C is an ideal with $(B \cap P)C \subseteq BP$ then $C \subseteq P$.

Suppose $C \not\subseteq P$, then $C \in \Phi_P$ and $0 = x(B \cap P)C \supseteq xPBC$. Now since $x \in E(R/P) \in F_P$, we have $xPB = 0$ and as in Claim 3 $x \in Q_i$ an impossibility.

Claim 5. If C is an ideal with $C(B \cap P) \subseteq BP$ then $C \subseteq B$.

Since $0 = xC(B \cap P) \supseteq xCPB$ we have that $xCP = 0$ so $xC \subseteq Q_i$ and by the maximality of B, $C \subseteq B$.

Now claims 4 and 5 yield that $B \cap P/BP$ is a left R/B right R/P bimodule and by Jategaonkar [74] Krull dim(R/B) = Krull dim(R/P) = Krull dim (R/P_j) an immpossibility by Jategaonkar [74] so $X_i = Q_i$ for all $i = 1,2,\ldots,n$. Thus E/Q is Φ_S-torsion free and the theorem follows by C - S [76].

Corollary 3.2. If P is a prime ideal of an FBN ring then P/P^n is right localizable in R/P^n for $n = 1,2,\ldots$.

We now show that the localizability of a prime ideal in an FBN ring is dependent only on the ideals of the ring.

Theorem 3.3. Let P be a prime ideal of an FBN ring R, then the following statements are equivalent.

1. P is right localizable.

2. If B is a prime ideal of R with $\kappa(R/B) = \kappa(R/P)$ and $B \neq P$, then there is an ideal $C \not\subseteq P$ such that $PC \subset BP$.

Proof. 1. implies 2.

Since $B \neq P$ and $\kappa(R/B) = \kappa(R/P)$ it follows by Jategaonkar [74] that $B \not\subseteq P$, hence $B \in \Phi_P$. By C - S [76] P/BP is Φ_P-torsion. Since R is left Noetherian $P = \sum_{i=1}^{n} Ra_i$. Now there exists $I_i \in \Phi_P$ such that $a_i I_i \subseteq BP$. Now $\bigcap_{i=1}^{n} I_i \in \Phi_P$ is contained in the annihilator of the right R-module P/BP so C the annihilator of the right R-module P/BP is in

Φ_P, hence $C \not\subseteq P$ and 2. follows.

2. implies 1. Let $t_P(E/Q) = X/Q$ where $E = E(R/P)$ and $Q = \mathrm{Ann}_E(P)$. We proceed to show that $X = Q$. Suppose $Q \subsetneq X$, then let B be an ideal that is maximal with respect to the property that there is an $x \in X - Q$ such that $xB \subseteq Q$.

If $\kappa(R/P) = \alpha$, then by Jategaonkar [74] E is an α-smooth module. Since the submodule of X/Q generated by $\bar{x} = x + Q$ is a finitely generated faithful right R/B module $\kappa(R/B) = \kappa(\bar{x}R)$. Now since E is α-smooth $\kappa(\bar{x}R) \leq \kappa(xR) = \alpha$ so it follows that $\kappa(R/B) \leq \kappa(R/P)$. Now it is not possible that $B = P$ since then $\bar{x}R$ would then be a finitely generated faithful singular R/P module which is impossible by Jategaonkar [74]. Therefore it follows that B and P are incomparable so $B \not\subseteq P$, hence $B \in \Phi_P$.

<u>Claim 1</u>. $x(B \cap P) \neq 0$.

If $0 = x(B \cap P) \supseteq (xP)B$, then $xP = 0$ since $B \in \Phi_P$ and therefore $x \in Q$ an impossibility.

<u>Claim 2</u>. $BP \subsetneq B \cap P$.

This is clear from Claim 1 as $xBP = 0$.

<u>Claim 3</u>. If C is an ideal such that $(B \cap P)C \subseteq BP$, then $C \subseteq P$.

<u>Proof</u>. If $C \not\subseteq P$, then $C \in \Phi_P$ and $0 = xBP \supseteq x(B \cap P)C$ so $x(B \cap P) = 0$ contrary to claim 1.

<u>Claim 4</u>. If C is an ideal such that $C(B \cap P) \subseteq BP$, then $C \subseteq B$.

<u>Proof</u>. $0 = xBP \supseteq xC(B \cap P) \supseteq xCPB$ so $xCP = 0$, hence $xC \subseteq Q$ and by the maximality of B, $C \subseteq B$.

Now by Claims 3, 4 $B \cap P/BP$ is a left R/B, right R/P bimodule which is faithful and finitely generated from the left and right. By Jategaonkar [74] $\kappa(R/B) = \kappa(R/P)$.

Now by 2 there is an ideal $C \not\subseteq P$ such that $PC \subseteq BP$ so $xPC = 0$, hence $xP = 0$ and therefore $x \in Q$ a contradiction so 1. follows.

Theorem 3.4. Let P be a prime ideal of an FBN ring R, then the following statements are equivalent.

1. $E = E^P$
2. If B is a prime ideal of R with $\kappa(R/B) = \kappa(R/P)$, $B \neq P$ and n a positive integer, then there exists an integer $m \geq n$ and an ideal $C \not\subseteq P$ such that

$$B \cap P^m \subseteq BP^nC$$

Proof. 1. implies 2. Since $B \in \Phi_P$ it follows by C - C [76] that there is an integer $m \geq n$ such that $B \cap P^m \subseteq \rho_P(BP^n)$. Since $(B \cap P^m)/BP^n = M_R$ is Φ_P-torsion and $B \cap P^m$ is finitely generated as a left ideal the annihilator of M_R, an ideal of R is in Φ_P and 2. follows.

2. implies 1. Suppose $E \neq E^P$. Let B be an ideal which is maximal such that there is an $x \in E - E^P$ with $xB \subseteq P$. Now from $\{x \in E - E^P | xB \subseteq E^P\}$ choose y so that n is the least positive integer so that $yB \subseteq E_n$. Clearly B is a prime ideal.

Claim 1. $\kappa(R/B) \leq \kappa(R/P)$.

Since the submodule of E/E^P generated by $y + E^P = \bar{y}$ is a finitely generated faithful right R/B module, $\kappa(R/B) = \kappa(\bar{y}R)$ by Jategaonkar [74]. Now since E is α-smooth where $\alpha = \kappa(R/P)$, $\kappa(R/P) = \alpha = \kappa(yR)$. Thus $\kappa(R/B) = \kappa(\bar{y}R) \leq \kappa(yR) = \kappa(R/P)$.

Claim 2. $B \not\subseteq P$.

If $B \subseteq P$ then by Claim 1 and Jategaonkar we would have $B = P$, hence $yP \subseteq E^P$, so $y \in E^P$, an impossibility.

Claim 3. $y(BP^{n-1} \cap P^n) \neq 0$ $(P^0 = R)$.

Suppose the claim is false, then $0 = y(BP^{n-1} \cap P^n) \supseteq yP^nBP^{n-1}$ and therefore $yP^nB \subseteq E_{n-1}$. By the minimality of n $yP^n \subseteq E^P$, hence $y \in E^P$ an impossibility.

Claim 4. If C is an ideal such that $(BP^{n-1} \cap P^n)C \subseteq BP^n$, then $C \subseteq P$. Suppose $C \not\subseteq P$, then $C \in \Phi_P$. Now $0 = y(BP^{n-1} \cap P^n)C$ so $y(BP^{n-1} \cap P^n) = 0$ by Claim 3 an impossibility.

Claim 5. If D is an ideal such that $D(BP^{n-1} \cap P^n) \subseteq BP^n$, then $D \subseteq B$. Suppose not, then $0 = yD(BP^{n-1} \cap P^n) \supseteq yDP^nBP^{n-1}$ and therefore $yDP^nB \subseteq E_{n-1}$. By the minimality of n $yDP^n \subseteq E^P$ and therefore $yD \subseteq E^P$ from which it follows that $D \subseteq B$.

Now by Claims 4 and 5 $(BP^{n-1} \cap P^n)/BP^n$ is left R/B right R/P faithful bimodule which is finitely generated on the left as well as the right. By Jategaonkar [74] $\kappa(R/B) = \kappa(R/P)$. Now by 2. there is an index $m \geq n$ and an ideal $C \not\subseteq P$ such that $B \cap P^m \subseteq BP^nC$ hence $0 = y(B \cap P^m) \supseteq yP^mB$ so $yP^m = 0$. This is an impossibility since $B \in \Phi_P$ and 1. follows.

Theorem 3.4. For P a prime ideal of an FBN ring R the following are equivalent.

1. P is right localizable.
2. $E(R/P)$ is P-primary.

Proof. 1. implies 2. This is a consequence of Corollary 3.6 in Jategaonkar [74] and may be found in Müller [75]. 2. implies 1. This follows from Corollary 3.2 and C - S [76, Theorems 3.7, 3.8]. For examples of localizable and nonlocalizable prime ideals in FBN rings see Müller [75].

Open Problems.

1. Is statement 2 of Theorem 3.4 a consequence of the same statement with n = 1?
2. Is Corollary 3.2 true when P is a semiprime ideal?
3. Is Corollary 3.2 true when R is a right fully bounded Noetherian ring?
4. In Theorem 3.3 statement 2 can C be chosen a prime ideal with $\kappa(R/C) = \kappa(R/P)$?
5. What are the FBN rings for which every prime ideal is right (left and right) localizable?

References

[60] H. Bass, "Finitistic Dimension and a Homological Generalization of Semiprimary Rings," Trans. Amer. Math. Soc. 95 (1960), 466-488.

[56] H. Cartan and S. Eilenberg, "Homological Algebra," Princeton University Press, 1956.

[76] J. H. Cozzens and F. L. Sandomierski, "Localization at a semiprime ideal of a right Noetherian ring," J. of Algebra (to appear).

[67] A. W. Goldie, "Localization in non-commutative Noetherian rings," J. Algebra (1967), 89-105.

[73] A. V. Jategaonkar, "The torsion theory at a semiprime Ideal," An. Acad. Brasil. Ciênc. 45 (1973), 197-200.

[73] ________, "Injective modules and classical localization in Noetherian rings," Bull. AMS 79 (1973), 152-157.

[74] ________, "Jacobson's conjective and modules over fully bounded Noetherian rings," J. of Algebra 30, 103-121.

[66] J. Lambek, "Lectures on Rings and Modules," Ginn-Blaisdell, 1966.

[73] _______ and G. O. Michler, "The torsion theory at a prime ideal in a right Noetherian ring," J. of Algebra 25 (1973), 364-389.

[74] _______, "Localization of right Noetherian rings at semiprime ideals," Can. J. Math. 5 (1974), 1069-1085.

[72] G. O. Michler, "Right symbolic powers and classical localization in right Noetherian rings," Math. Zeit. 127 (1972), 57-69.

[70] K. Morita, "Localization in Categories of Modules I," Math. Z. 114 (1970), 121-144.

[74] B. Mueller, "Localization of noncommutative Noetherian rings at semiprime ideals," McMaster Univ. publications.

[76] B. J. Müller, "Localization in non-commutative Noetherian rings," Can. J. Math. Vol. 28 (3), (1976), 600-610.

[75] _______, "Localization in Fully Bounded Noetherian Rings," Math. Report. No. 78, McMaster Univ. Publ. Dec. 1975.

SOME DECOMPOSITION THEOREMS IN ABELIAN GROUPS AND THEIR GENERALIZATIONS

Surjeet Singh*

Guru Nanak Dev University
Amritsar, India

Basic contributions to the problem of decomposability of abelian p-groups were made by Kulikov, Kertesz, Prüfer, and Szele. Kaplansky [3] studied analogous problems for modules over (commutative) Dedekind domains. Marubayashi generalized some of these results to torsion modules over bounded Dedekind prime rings and in a recent paper [7] we have done analogous work for modules over bounded (hnp)-rings. The main purpose of this note is to point out that large number of results on decomposability of torsion abelian groups, and which have been generalized to modules over various types of rings mentioned above, do not depend upon the nature of the ring involved; rather they are characteristics of modules with some properties. Since our intention is only to point out the way the results can be generalized, we shall prove only a few results. An application of these results to (hnp)-rings is given in Theorem 6.

Since proofs of many results here, are verbatim same as for analogous results for torsion modules over bounded (hnp)-rings proved in [7], in all such cases, either the results will be stated without proof or else will be directly referred to [7]. The notations in [7] will be used essentially without comments.

---

*This research was supported by U.G.C. Grant No. F 30-5(6562)/76(SR-II).

Consider the following two conditions on a module M over a ring R:

(I) Every finitely generated submodule of every homomorphic image of M is a direct sum of uniserial modules.

(II) Given two uniserial submodules U and V of a homomorphic image of M, for any submodule W of U, any non-zero homomorphism $f:W \to V$ can be extended to a homomorphism $g:U \to V$ provided composition length $d(U/W) \leq d(V/f(W))$.

Throughout let M be an R-module satisfying (I) and (II). An element $x\ (\neq 0) \in M$ is called a uniform element if xR is a uniform submodule (hence uniserial by (I)). Given a uniform element $x \in M$, d(xR) is called the exponent of x, denoted by e(x); the $\sup\{d(T/xR)\}$ where T runs over all those uniserial submodules of M, which contain x, is called the height of x in M, and is denoted by $H_M(x)$, or some-times simply by H(x). These concepts of exponent and height generalize the analogous concepts for abelian p-groups, where p is any prime.

It can be seen easily that the proofs of Lemmas 3, 4, 5, 6 in [7] only depend upon Lemmas 1 and 2(b) in [7]. A module satisfying conditions (I) and (II) satisfies the conclusions in the lemmas mentioned above. Hence Lemmas 3, 4, 5, 6 in [7] and as a result Theorem 3 in [7] hold for a module M satisfying (I) and (II).

Thus if for any $k \geq 0$, $H_k(M)$ denotes the submodule of M generated by all the uniform elements of height at least k, then every uniform element in $H_k(M)$ is of height at least k. Further we obtain the following generalization of Kulikov's theorem [2, Theorem (11.1)].

Theorem 1. Let M be an R-module satisfying (I) and (II). M is a direct sum of uniserial R-submodules if and only if M is a union of ascending sequence $M_n (n=1,2,3,\ldots)$ of submodules such that for each n, there exists a positive integer k_n, with the property: $H_M(x) \leq k_n$ for all uniform elements x of M_n.

M is said to be bounded if there exists a positive integer k such that $H(x) \leq k$ for all uniform elements x in M. So the following is immediate from Theorem 1.

Corollary 1. Any bounded module satisfying (I) and (II) is a direct sum of uniserial modules.

Again Lemma 7 in [7] also holds for M (Proof being same). As a result we get the following theorem which generalizes [2, Theorem (24.1)].

Theorem 2. Let M be a module satisfying (I) and (II) and let N be a submodule of M such that N is a direct sum of uniserial modules of same length k. Then the following are equivalent:

(i) N is a summand of M.

(ii) $H_n(N) = N \cap H_n(M)$ for all n.

(iii) N satisfies $H_k(M) \cap N = 0$.

As a consequence we get the following result which generalizes [7, Theorem 8].

Theorem 3. Let M be a module satisfying (I) and (II). If a submodule N of M is bounded and satisfies $H_n(N) = N \cap H_n(M)$ for all n, then N is a summand of M.

The above theorem motivates the following:

Definition 1. A submodule N of a module M-satisfying (I) and (II) is called an h-pure submodule of M, if $H_n(N) = N \cap H_n(M)$.

For torsion modules over a bounded (hnp)-ring, the above concept has been recently shown to be equivalent to that of pure submodules [8].

It follows from (I) that socle (M) is large in M.

Lemma 1. Let x be a uniform element in socle (M) such that $H(x)$ is finite. If $u \in M$ is a uniform element such that $x \in uR$ and $d(uR/xR) = H(x)$, then uR is a h-pure submodule of M and hence a summand of M.

Proof. For any $z(\neq 0) \in uR$, $H_{uR}(z) = n+1 - e(z)$. Also as $x \in zR$ and $H(x) = n$, $H_M(z) \leq n - d(zR/xR) = n - e(z) + 1$. Hence $H_M(z) = H_{uR}(z)$. Consequently uR is h-pure. Hence by Theorem 3, uR is a summand of M.

Lemma 2. Let N be a submodule of M. The following hold:

(i) If N is h-pure in M, given any uniform element $\bar{x} \in M/N$, there exists a uniform element $x' \in M$ such that $e(\bar{x}) = e(x')$ and $\bar{x} = \bar{x}'$.

(ii) If N is h-pure in M, and K is any submodule of N, N/K is h-pure in M/K.

(iii) If K is an h-pure submodule of M such that $K \subset N$, and N/K is h-pure in M/K, then N is h-pure in M.

Proof (i). Using condition (I) we can find a uniform element z in M of smallest exponent such that $\bar{x} = \bar{z}$. If $zR \cap N = 0$, take $x' = z$. Otherwise let $yR = zR \cap N \neq 0$. Then $d(zR/yR) = d(\bar{x}R) = n$. As $N \cap H_n(M) = H_n(N)$, we can find a uniform element $z' \in N$ such that $y \in z'R$ and $d(z'R/yR) = n$. Using condition (II) we get an isomorphism $\eta : zR \to z'R$, which is identity on yR and we can take $z' = \eta(z)$. Then we have natural epimorphism $zR \to (z - z')R$ whose kernel contains yR. Hence $e(z - z') \leq d(zR/yR) = d(\bar{x}R) = e(\bar{x})$. As $\bar{x} = \overline{z - z'}$ we have $e(\bar{x}) \leq e(z - z')$, consequently $z - z'$ is a desired uniform element.

Proofs of (ii) and (iii) can be very well adapted from those for pure subgroups.

Lemma 3. Let every uniform element in socle (M) be of infinite height. Then given any uniform element $x \in M$, there exists a uniform element $y \in M$ such that $x \in yR$ and $e(x) < e(y)$.

Proof. Same as in the first two paragraphs of proof of Lemma 8 in [7].

Theorem 4. Let M satisfy (I) and (II). If every uniform element in socle (M) is of infinite height, then M is a direct sum of infinite length uniform submodules.

Proof. By condition (I) every uniform submodule of M is either uniserial (hence cyclic) or is of infinite length, in which every proper submodule is of finite length. Lemma 3 shows that every maximal uniform submodule of M is of infinite length. If N is a maximal uniform submodule of M, of infinite length, then the height in N of any uniform element of N, is infinite. So that the equality $H_n(N) = M \cap H_n(M)$ for all n holds trivially; thus N is an h-pure submodule of M. Thus as union of any chain of h-pure submodules of M is h-pure, we can find a maximal direct sum $M' = \oplus \Sigma N_\alpha$ of infinite length uniform submodules such that M' is an h-pure submodule of M. Suppose $M \neq M'$, using Lemma 2(i) we can find a uniform element x in socle (M) such that $x \notin M'$ and so $xR \cap M' = 0$. By Lemma 3, we get an infinite length uniform submodule N of M such that $x \in N$. Then $N \simeq \frac{N + M'}{M'}$, which as remarked earlier is an h-pure submodule of M/M'. Hence by Lemma 2(iii), N + M' is an h-pure submodule of M clearly $N \cap M' = (0)$, so $N + M' = N \oplus M'$. This violates the maximality of M'. Consequently $M = M'$. This proves the theorem.

The following theorem generalizes a wellknown result for torsion abelian groups.

Theorem 5. Let a module M satisfy (I) and (II). Then M has uniform summand; which can be chosen to be of finite length in case not all uniform elements in socle (M) are of infinite height.

Proof. Immediate from Theorem 4 and Lemma 1.

We end this note by giving an application to (hnp)-ring. Eisenbud and Griffith [1], proved that given any (hnp)-ring R, R/A is a generalized uniserial ring for any nonzero ideal A. A partial converse of the above result was proved by Zaks [9]. He showed that if R is a prime, noetherian (on both sides), bounded ring, whose every proper homomorphic image is Quasi-Frobenius (hence a uniserial ring), then R is a Dedekind prime ring. We improve this result and prove the following.

Theorem 6. Let R be a prime right Goldie, right bounded ring such that for every non-zero ideal A of R, R/A is a generalized uniserial ring, then R is right hereditary.

Proof. Since R is right bounded every finitely generated torsion right R-module M has non zero annihilator A. So M is a module over a generalized uniserial ring and hence a direct sum of uniserial modules. Thus any torsion R-module satisfies condition (I). In this case any injective uniform torsion right R-module is such that family of all its submodules is totally ordered and its every proper submodule is of finite length. This fact yields that any torsion right R-module satisfies (II). Using (II) we get that any uniform torsion R-module is either uniserial, or injective and of infinite length. Notice also that R is right noetherian.

Let N be any divisible (right) R-module and T be its torsion submodule. T is also divisible. If T is non-zero, then is faithful; so it cannot have finite length direct summands. Consequently Theorem 5 and the discussion in the previous paragraph give that T is injective. Thus in any case $N = S \oplus T$, where S is torsion free. Hence S being also divisible is injective by [4]. Consequently N is injective. That proves that every divisible right R-module is injective and by Levy [4], R is right hereditary. This proves the theorem.

References

[1] D. Eisenbud and P. Griffith, "Serial rings," J. Algebra, 17 (1971), 389-400.

[2] L. Fuchs, "Abelian Groups," Pergamon Press, New York, 1960.

[3] I. Kaplansky, "Modules over Dedekind rings and valuation rings," Trans. Amer. Math. Soc. 72 (1952), 327-340.

[4] L. S. Levy, "Torsion free and divisible modules over non-integral domains," Can. J. Math. 15 (1963), 132-151.

[5] H. Marubayashi, "Modules over bounded Dedekind prime rings," Osaka J. Math., 9 (1972), 95-110.

[6] ________, "Modules over bounded Dedekind prime rings II," Osaka J. Math., 9 (1972), 427-445.

[7] S. Singh, "Modules over hereditary noetherian prime rings," Can. J. Math., 27 (1975), 867-883.

[8] ________ and M. Khan, "On Pure submodules," (submitted).

[9] A. Zaks, "Some rings are hereditary rings," Israel J. Math., 10 (1971), 442-450.

IDENTITIES AND NILPOTENT ELEMENTS IN

LATTICE-ORDERED RINGS

Stuart A. Steinberg

University of Toledo
Toledo, Ohio

1. Introduction

A lattice-ordered ring (ℓ-ring) is a ring R which is also a lattice in which the ring and lattice structures interact via the conditions:

$$a \le b \text{ implies } a + c \le b + c \text{ and}$$

$$0 \le a \text{ and } 0 \le b \text{ implies } 0 \le ab$$

for all elements a, b and c of R. Throughout this paper R will denote an ℓ-ring.

An ℓ-ring is called an f-ring if it is embeddable in a product of totally ordered rings. The class of f-rings was introduced by Birkhoff and Pierce in [2] where it was shown to be a variety determined by a single identity. It is the most extensively studied class of ℓ-rings for the obvious reason that one can readily compute in totally ordered rings.

It is natural to obtain reasonable classes of ℓ-rings which properly contain the variety of f-rings by considering identities satisfied by every totally ordered ring. Two such identities singled out by Birkhoff and Pierce [2, p. 57] are $x^+x^- = 0$ and $(x^2)^- = 0$. (If x is an element of R, then the positive part, the negative part and the absolute value of x are defined by $x^+ = x \vee 0$, $x^- = (-x) \vee 0$ and

$|x| = x \vee (-x)$, respectively.) In [2, Theorem 15, p. 60] Birkhoff and Pierce prove that an ℓ-ring with a unit element satisfies the identity $x^+x^- = 0$ if and only if the unit element is a weak order unit ($a \in R$ is a weak order unit if $a \geq 0$ and $a \wedge x = 0$ implies $x = 0$). The main result of this paper is the following generalization of this theorem.

Theorem 8. Let e be an f-element and a left superunit in the ℓ-ring R and let S be the convex ℓ-subring of R generated by e. Consider the following statements:

(a) R satisfies $x^+x^- = 0$.

(b) e is a weak order unit of R.

(c) If $a \geq 0$ and $a \wedge e$ is nilpotent, then $a \in S$.

(d) If $a \geq 0$ and $(a \wedge e)^2 = 0$, then $a \in S$.

(e) R satisfies $(x^2)^- = 0$ and if $a \geq 0$ with $(a \wedge e)^2 = 0$, then $a^2 = 0$.

(f) R is a P.P.I. ℓ-ring and if $a \geq 0$ with $(a \wedge e)^2 = 0$, then $a^2 = 0$.

Then (a) through (e) are equivalent, (e) implies (f), and if $e = 1$, (f) is equivalent to (a).

The proof of this result is modelled after the arguments of Birkhoff and Pierce and is given in section 3. For the convenience of the reader we give an appreciable amount of background material on lattice-ordered rings in section 2. In section 4 there are some examples and additional remarks while in section 5 there appears a generalization of a theorem from [13].

2. Background

We first collect some basic definitions and properties of ℓ-rings; see [2], [6] and [3]. The positive cone of the ℓ-ring R is

$R^+ = \{x \in R : x \geq 0\}$. An <u>$\ell$-group</u> may be conveniently defined as an ℓ-ring with trivial multiplication: $xy = 0$ for each x and y. A <u>convex ℓ-subgroup</u> of the ℓ-ring R is a subgroup A of the additive group of R which is convex, i.e. $a \leq b \leq c$ with $a, c \in A$ implies $b \in A$, and which is also a sublattice. An <u>ℓ-ideal</u> of R is an ideal which is also a convex ℓ-subgroup. The ℓ-ideals of R are precisely the kernels of homomorphisms: a homomorphism between two ℓ-rings is a mapping which is a ring and lattice homomorphism. If A is an ℓ-ideal of R, then the relation $x + A \leq y + A$ if $x \leq y + a$ for some $a \in A$ is a partial order of R/A. If X is a subset of R, then the <u>polar</u> of X is

$$X' = \{y \in R : |y| \wedge |x| = 0 \text{ for each } x \in X\}.$$

The proof of the following theorem may be found in [2, chapters 13 and 17] and [6, chapters 5 and 9].

<u>Theorem 1</u>. Let x, y and z be elements of the ℓ-ring R and let n be a positive integer.

(a) $x + y = (x \vee y) + (x \wedge y)$.

(b) If $nx \geq 0$, then $x \geq 0$.

(c) The underlying group of R is torsion-free.

(d) The underlying lattice of R is distributive.

(e) If $x \wedge y = 0$, then $x = z^+$ and $y = z^-$ for some element z of R.

(f) Each polar is a convex ℓ-subgroup.

(g) $|x| \geq 0$ and $|x| = 0$ if and only if $x = 0$.

(h) $x^+ \wedge x^- = 0$.

(i) $x = x^+ - x^-$ and $|x| = x^+ + x^-$.

(j) $n(x \vee y) = nx \vee ny$.

(k) $|x + y| \leq |x| + |y|$.

(l) If $x, y, z \in R^+$ and $x \leq y + z$, then $x = y_1 + z_1$

where $0 \leq y_1 \leq y$ and $0 \leq z_1 \leq z$

(m) If A is an ℓ-ideal of R, then R/A with the partial order defined above is an ℓ-ring.

(n) $|xy| \leq |x||y|$.

(o) The set of ℓ-ideals of R is a distributive sublattice of the lattice of ideals of R.

(p) The ℓ-ideal generated by x is $I(x) = \{y \in R: |y| \leq m|x| + r|x| + |x|s + t|x|u$ where $r, s, t, u \in R$ and m is an integer$\}$.

Theorem 2 [3, Theorem 12, p. 57]. The following statements are equivalent for an ℓ-ring R.

(a) R is an f-ring.

(b) R satisfies the identity $(x^+y^+ \wedge x^-) \vee (y^+x^+ \wedge x^-) = 0$.

(c) If a, b, c $\in R^+$ and $a \wedge b = 0$, then $ac \wedge b = ca \wedge b = 0$.

(d) Each polar of R is an ℓ-ideal.

(e) Each subdirectly irreducible homomorphic image of R is totally ordered.

Proof. If R is totally ordered, then for each $x \in R$, $x^+ = 0$ or $x^- = 0$. So the identity in (b) holds in any totally ordered ring and hence in any f-ring. Thus (a) implies (b).

By (e) of Theorem 1 (b) and (c) are equivalent. Since each element of R is a difference of positive elements and since a polar is a convex ℓ-subgroup (Theorem 1 (i) and (f)) (c) just says that a principal polar is an ℓ-ideal. But then (c) and (d) are equivalent since each polar is clearly an intersection of principal polars.

Let R be a subdirectly irreducible ℓ-ring which satisfies (b) and let a, b $\in R^+$ with $a \wedge b = 0$. If $0 \leq x \in I(a) \cap I(b)$, then by (p) of Theorem 1

$$x \leq na + ra + as + tau = c \quad \text{and}$$

$$x \leq mb + vb + bw + ybz = d.$$

Since b' is an ℓ-ideal and $a \in b'$, we must have $c \in b'$, i.e. $b \wedge c = 0$. Similarly we must have $d \in c'$ since $b \in c'$. So $0 \leq x \leq c \wedge d = 0$. Thus $I(a) \cap I(b) = 0$ and since R is subdirectly irreducible $a = 0$ or $b = 0$. But then R is totally ordered by (h) of Theorem 1.

In addition to the identities $x^+x^- = 0$ and $(x^2)^- = 0$ another important identity satisfied by any f-ring is $|xy| = |x||y|$. An ℓ-ring satisfying this last identity is usually called a <u>distributive</u> ℓ-ring or <u>d-ring</u>.

<u>Theorem 3</u> [3, pp. 58-60]. (a) If R satisfies $x^+x^- = 0$, then it satisfies $(x^2)^- = 0$, i.e. the square of every element of R is positive.

(b) R is a d-ring if and only if it satisfies the identities $(x \vee y)z^+ = xz^+ \vee yz^+$ and $z^+(x \vee y) = z^+x \vee z^+y$.

(c) If R is a d-ring in which $xR = 0$ or $Rx = 0$ implies $x = 0$ for each $x \in R^+$, then R is an f-ring.

(d) A d-ring which is either commutative or satisfies $(x^2)^- = 0$ must satisfy $x^+x^- = 0$.

<u>Proof.</u> (a) $x^2 = (x^+ - x^-)^2 = (x^+)^2 + (x^-)^2 \geq 0$ since $x^+x^- = x^-x^+ = 0$ (Theorem 1 (i), (h) and (e)).

(b) Suppose that R is a d-ring and $c \in R^+$. Then $x \vee y = [(x - y) \vee 0] + y$ so $(x \vee y)c = (x - y)^+c + yc$. By (i) of Theorem 1 $(x - y) + |x - y| = 2(x - y)^+$, so $2[(x - y)c]^+ = (x - y)c + |(x - y)c| = (x - y)c + |x - y|c = 2(x - y)^+c$. Thus $[(x - y)c]^+ = (x - y)^+c$ and $(x \vee y)c = (xc - yc)^+ + yc = xc \vee yc$. Similarly $c(x \vee y) = cx \vee cy$.

Conversely, suppose that R satisfies these two identities. Then $|xy^+| \wedge |xy^-| = |x|y^+ \wedge |x|y^- = |x|\ (y^+ \wedge y^-) = 0$; so

$$|x||y| = |x|(y^+ + y^-) = |xy^+| + |xy^-| = |xy^+| + |-xy^-| =$$
$$|xy^+ - xy^-| = |xy|.$$

The next to the last equality comes from the fact that if $|a| \wedge |b| = 0$, then $|a + b| = |a| + |b|$. For this is certainly true in a totally ordered group. But by Theorem 2 the underlying ℓ-group of R is a subdirect product of totally ordered groups and so it is true in R.

(c) Suppose that $a \wedge b = 0$ and $c \geq 0$. Then $0 \leq (ac \wedge b)r = acr \wedge br \leq a(cr \vee r) \wedge b(cr \vee r) = (a \wedge b)(cr \vee r) = 0$ for each $r \wedge R^+$. So $ac \wedge b = 0$ and similarly $ca \wedge b = 0$.

(d) If R is a commutative d-ring and $a \wedge b = 0$, then $0 = (a \vee b)(a \wedge b) = [(a \vee b)a] \wedge [(a \vee b)b] = (a^2 \vee ba) \wedge (ab \vee b^2) = ab \vee (a^2 \wedge b^2) \geq ab = 0$.

If R is a d-ring in which the square of every element is positive and $a, b \in R^+$, then $0 \leq (a - b)^2$ yields $ab \leq a^2 + b^2$. So if $a \wedge b = 0$, then

$$0 \leq ab = ab \wedge (a^2 + b^2) \leq (ab \wedge a^2) + (ab \wedge b^2) =$$
$$a(b \wedge a) + (a \wedge b)b = 0.$$

Note that one cannot replace "commutative" by "a P.I. ring" in (d). For if $R = \{\begin{pmatrix} a & b \\ o & o \end{pmatrix}$: a, b rationals$\}$ and $R^+ = \{\begin{pmatrix} a & b \\ o & o \end{pmatrix}$: $a \geq 0$ and $b \geq 0\}$, then (R, R^+) is a d-ring which is a P.I. ring but which does not satisfy $x^+x^- = 0$.

A radical theory for ℓ-rings can be developed just as for rings [5]. So if P is a radical class of ℓ-rings and R is an ℓ-ring, then P(R) is the largest ℓ-ideal of R in P and R/P(R) has no nonzero ℓ-ideals in P. In particular we denote the analogues of the lower radical, Levitzki radical, nil radical and generalized nil radical by ℓ-β, ℓ-Lev, ℓ-Nil and ℓ-Ng, respectively. (R = ℓ-β(R) if every nonzero homomorphic image of R has a nonzero nilpotent ℓ-ideal; R = ℓ-Lev(R) if every finitely generated ℓ-subring of R is nilpotent; R = ℓ-Nil(R) if R is nil and R = ℓ-Ng(R) if every nonzero homomorphic image of R has a nonzero positive nilpotent element.) Each of these ℓ-radicals has characterizations similar to those of its corresponding ring radical (see [4], [10], [11] and [14]). Here we merely list the requisite definitions.

An ℓ-ring R is called <u>ℓ-semiprime</u> if it does not contain any nonzero nilpotent ℓ-ideals. It is <u>ℓ-prime</u> if whenever I and J are ℓ-ideals with IJ = 0, then either I=0 or J=0. An ℓ-ideal I is <u>ℓ-prime</u> if R/I is an ℓ-prime ℓ-ring. R is called an <u>ℓ-domain</u> if a, b $\in R^+$ and ab = 0 imply that a = 0 or b = 0. If X is a nonempty subset of R, then the <u>left ℓ-annihilator of</u> X is $\ell_\ell(X) = \{r \in R: |r||x| = 0$ for each $x \in X\}$. Finally, X is an <u>ℓ-m-system of</u> R if $X \subseteq R^+$ and for each x, y $\in$ X there is an element z $\in R^+$ with xzy $\in$ X.

In contrast to the ring theoretic situation Shatalova showed that the ℓ-radicals ℓ-Lev and ℓ-Nil are identical. In fact, he showed more.

<u>Lemma 1</u> [11, Lemma 5]. Let S be the ℓ-subring of R generated by $X = \{x_1, \ldots, x_n\}$. Then the following statements are equivalent.

(a) S is nilpotent.

(b) $\bigvee_{i=1}^{n} |x_i|$ is nil.

(c) There is an integer N with $|x_{i_1}| \ldots |x_{i_N}| = 0$ for all $x_{i_1}, \ldots, x_{i_N} \in X$.

<u>Proof</u>. Clearly (a) implies (b), and if $x = \bigvee |x_i|$ with $x^N = 0$, then $0 \leq |x_{i_1}| \ldots |x_{i_N}| \leq x^N = 0$; so (b) implies (c).

To see that (a) is a consequence of (c) let us denote by L(Y) the sublattice of R generated by the subset Y and by R(Y) the subring generated by Y. Then put $R_1(X) = R(X)$, $L_1(X) = L(X)$, $R_{k+1}(X) =$

$R(L_k(X))$ and $L_{k+1}(X) = L(R_{k+1}(X))$. So $S = \bigcup_k R_k(X)$. By induction we show that $|p_1|\dots|p_N| = 0$ if $p_1,\dots,p_N \in R_k$.

If $p_1,\dots,p_N \in R_1(X)$, then $0 \leq |p_1\dots p_N| \leq |p_1|\dots|p_N| \leq \Sigma|x_{i_1}| \dots |x_{i_N}| = 0$. Suppose $\ell_1,\dots,\ell_N \in L(R_k(X)) = L_K(X)$. Since R is a distributive lattice $\ell_\nu = \wedge_I \vee_J a_{ij}^{(\nu)}$ where $a_{ij}^{(\nu)} \in R_k(X)$ for some finite sets I and J. Now $|\ell_\nu| \leq \Sigma_{i,j} |a_{ij}^{(\nu)}|$ since this is true in any totally ordered homomorphic image of the additive group of R (Theorem 2). But then $|\ell_1|\dots|\ell_N| \leq \Sigma|b_{i_1}|\dots|b_{i_N}|$ where $b_{i_\tau} \in R_k(X)$, so $|\ell_1|\dots|\ell_N| = 0$. Now if $p_1,\dots,p_N \in R_{k+1}(X)$, then since the p's are polynomials in the ℓ's, $|p_1|\dots|p_N|$ is dominated by a sum of monomials, each of which is a product of at least N elements from $L_k(X)$.

Theorem 4 [11, Corollary and Theorem 1]. For an ℓ-ring R the following statements are equivalent.

(a) R is nil.

(b) R is locally nilpotent.

(c) R is ℓ-locally nilpotent.

For f-rings a much stronger statement can be made.

Theorem 5 [3, p. 63]. If R is an f-ring, then $Z_n(R) = \{x \in R: x^n = 0\}$ is a nilpotent ℓ-ideal of R.

Proof. Since $|x_1\dots x_n| \leq (\vee|x_i|)^n$ in any ℓ-ring it is clear that an ℓ-ring which satisfies $x^n = 0$ is nilpotent of index at most n. So it suffices to show that $Z_n(R)$ is an ℓ-ideal of R. We may assume that R is totally ordered. If $x,y \in Z_n(R)$ and $|x| \leq |y|$, then $|x - y|^n \leq (2|y|)^n = 0$. And if $a \in R$ with $|ax| \leq |xa|$, then $|ax|^n \leq |xa|^n = (|x||a|)^n = |x|(|a||x|)^{n-1} |a| \leq \dots \leq |x|^n|a|^n = 0$.

So for an f-ring ℓ-$\beta(R)$ is the set of nilpotent elements of R. If R is a d-ring this is still true since by Theorem 3(c) R/ℓ-$\beta(R)$ is an ℓ-semiprime f-ring. As a consequence of the next theorem it is true also if R satisfies $x^+x^- = 0$. This theorem was proven by Diem [4, 3.6] but only stated for an ℓ-ring which satisfies $(x^2)^- = 0$. In such an ℓ-ring (b) is always satisfied since $0 \leq ab \leq a^2 + b^2$ if $a, b \geq 0$.

Theorem 6 [Diem]. An ℓ-ring R is an ℓ-domain if and only if it is ℓ-prime and satisfies (a) and (b):

(a) $a \wedge b = 0$ and $ab = 0$ imply $ba = 0$;

(b) $a^2 = b^2 = 0$ implies $ab = 0$ if $a, b \in R^+$.

Proof. Suppose R is ℓ-prime and satisfies the conditions and let $a, z \in R^+$ with $a^2 = 0$. We show that $aza = 0$ and so $a = 0$ since R is ℓ-prime; but then if $a, z, b \in R^+$ with $ba = 0$, $(azb)^2 = 0$ and so $aRb = 0$. Consequently $a = 0$ or $b = 0$ since R is ℓ-prime.

If $az \leq za$ or $za \leq az$ then clearly $aza = 0$. Otherwise $(za - az)^+ > 0$ and $(za - az)^- > 0$. Now $0 \leq (za - az)^+ \cdot (za - az)^- \leq (za)(az) = 0$, so by (a) $(za - az)^-(za - az)^+ = 0$. If $y \in R^+$, then $(za - az)^+y\,(za - az)^- \in R^+$ is nilpotent of index at most two, so by (b) $a(za - az)^+y\,(za - az)^- = 0$. Thus $a(za - az)^+ = 0$ and similarly $a(za - az)^- = 0$. So $0 = a(za - az) = aza$.

Corollary 1 [4, 4.3]. An ℓ-prime ℓ-ring is a totally ordered domain if and only if it satisfies $x^+x^- = 0$.

There is a question related to this corollary which was first asked by Diem. Must an ℓ-prime ℓ-ring with squares positive be an ℓ-domain? The answer is unknown but we present some evidence to suggest that the answer is yes (at least if $1 \in R$).

If M is an ℓ-module over the ℓ-ring R (i.e. M is an ℓ-group, an R-module, and $M^+R^+ \subseteq M^+$ - if M is a right R-module), an element $a \in R^+$ is called an f-element (d-element) on M provided the endomorphism of M induced by a preserves polars (is a lattice homomorphism). If R^+ consists of f-elements (d-elements) on M, then M is called an f-module (d-module). An f-element (d-element) on the ℓ-module R_R will be referred to as a right f-element (d-element) of R, and R is called a right f-ring (d-ring) if R_R is a f-module (d-module). Similar definitions can be made for ${}_R R$ and the adjectives left and right will be dropped if they both apply.

Theorems 2 & 3 hold for ℓ-modules [12]; and so an ℓ-module is an f-module exactly when it is embeddable in a product of totally ordered R-modules, and a right d-module is an f-module provided $\{x \in M^+ : xR^+ = 0\} = 0$. Let $S = \{x \in R : |x| \leq \sum_j n_j a^j,\ n_j \text{ an integer}\}$ be the convex ℓ-subring of R generated by $a \in R^+$; then it is easy to see that a is an f-element (d-element) on M if and only if M is an f-module (d-module) over S.

We next make some definitions in order to generalize the condition "the square of every element is positive." An ℓ-ring R is called an ℓ-algebra over the commutative partially ordered ring F if R is an algebra over F and an f-module over F. R is called a P.P.I. ℓ-algebra if there is a nonzero polynomial $p(x) \in F[x]$ with $p(a) \geq 0$ for each $a \in R$. If $1 \in R$, then we also require $p(1)p'(1) \neq 0$ and $p(1) - p(0) \neq 0$ ($p'(x)$ is the derivative of $p(x)$).

Diem [4, 3.3] proved that the nilpotent elements in an ℓ-ring with 1 which satisfies $x^2 \geq 0$ are all bounded by 1.

Theorem 7 [14, Theorem 2]. Let R be a torsion-free P.P.I. ℓ-algebra over the totally ordered domain F, and let S be the convex ℓ-subalgebra of R generated by 1. Then S contains all the nilpotent and idempotent elements of R and the latter are all central.

Proof. (a) We first note that $1 \in R^+$ and hence that S is an f-subalgebra of R. For $p(1)p'(1) \neq 0$ implies the existence of $\beta \in F$ with $\beta \cdot 1 > 0$. Since F is totally ordered and $-1 \notin R^+$ we must have $1 \in R^+$.

To see that every nilpotent element a is in S we argue by induction on the index of nilpotency n of a. The case $n = 1$ is trivial so we assume $n \geq 2$. Then $a^k \in S$ for each $k \geq 2$, since the index of a^k is at most $n - 2$. By replacing x by x^2 in $p(x)$ we may assume that the nonzero terms in $p(x)$ are all even. Let $p(x) = \alpha_0 + \alpha_2 x^2 + \ldots + \alpha_m x^m$. Then $0 \leq p(a-1) = \alpha_0 + \alpha_2(a-1)^2 + \ldots + \alpha_m(a-1)^m$ yields $0 \leq \beta_0 - p'(1)a + \beta_2 a^2 + \ldots + \beta_m a^m$ for some $\beta_i \in F$. Similarly, $p(a+1) \geq 0$ yields $0 \leq \gamma_0 + p'(1)a + \ldots + \gamma_m a^m$ for some $\gamma_i \in F$. So $-(\gamma_0 + \gamma_2 a^2 + \ldots + \gamma_m a^m) \leq p'(1)a \leq \beta_0 + \beta_2 a^2 + \ldots + \beta_m a^m$ implies that $p'(1)a \in S$ and hence that $\beta|a| \in S$ where $\beta = |p'(1)| > 0$. Since R is a torsion-free ℓ-algebra over F, the F-divisible hull R_1 of R is an ℓ-algebra over the totally ordered quotient field Q of F and R_1 contains R. Then $|a| \leq \alpha\beta^{-1}$ in R_1 for some $\alpha \in F$. But the convex ℓ-Q-subalgebra of R_1 generated by 1 is an f-subalgebra and so $|a| \leq 1$.

If e is an idempotent, then $0 \leq p(1-e) = \alpha_0 + (\alpha_2 + \ldots + \alpha_m)(1-e) = p(1) - (\alpha_2 + \ldots + \alpha_m)e$. Thus $(\alpha_2 + \ldots + \alpha_m)e \leq p(1)$. But also $0 \leq p(e) = \alpha_0 + (\alpha_2 + \ldots + \alpha_m)e$ so $(p(1)-p(0))e = (\alpha_2 + \ldots + \alpha_m)e \in S$. Thus, as above, $e \in S$.

Now the idempotent elements in an f-ring with unit are central [7, 2.1]. For if R is totally ordered and $e^2 = e$, then $e \leq 1-e$ or $1-e \leq e$. Hence $e = 0$ or $e = 1$ and so is central. Since an f-ring is a subdirect product of totally ordered rings its idempotents are central. Now S is an f-ring and so the idempotents of R commute and hence they are all central.

Note that if e is an idempotent in the P.P.I. ℓ-algebra R (with unit) and $|x| \leq ye$, then $|x(1-e)| \leq |x|(1-e) \leq ye(1-e) = 0$. So $x \in Re$ and Re is an ℓ-ideal. In particular, an ℓ-prime P.P.I. ℓ-algebra with 1 has only two idempotents.

Note also that an ℓ-domain with 1 and squares positive is, in fact, a domain. For if $ab = 0$, then $a^2b^2 = 0$ and so $a^2 = 0$ or $b^2 = 0$. But then a or b is a nilpotent element in the totally ordered domain S (Theorem 7 and Corollary 1) and so a or $b = 0$.

Another fact related to Diem's question is that an ℓ-prime ℓ-ring with squares positive is an ℓ-domain provided it contains no right ℓ-quasiregular right ℓ-ideals [14, Corollary 2]. (A right ℓ-ideal I is <u>right ℓ-quasiregular</u> if for each $0 \leq a \in I$ the right ℓ-ideal generated by $(1-a)R$ contains a.)

3. Proof of Theorem 8

A positive element e of the ℓ-ring R is called a <u>left(right) superunit</u> if $ex \geq 0$ ($xe \geq 0$) for each $x \in R^+$. An element which is both a left and a right superunit is called a <u>superunit</u>. The concept of a superunit was introduced by Henriksen and Isbell in [7], where it was used to characterize those f-rings that can be embedded in f-rings with units.

Throughout this section S will be the convex ℓ-subring of R generated e, and e will be a left superunit in S.

<u>Lemma 2</u>. Suppose that e is a weak order unit of R and S is a right f-ring. If $a \in R^+$ and $a \wedge e$ is nilpotent, then $a \leq e$.

<u>Proof</u>. Suppose that $a \wedge 2e \not\leq e$. Then, since S is a right f-ring, there is a nonzero totally ordered homomorphic image $\overline{S}_S$ of S_S in which $\overline{a \wedge 2e} > \overline{e}$. So $\overline{a \wedge e} = \overline{a \wedge 2e \wedge e} = \overline{a \wedge 2e} \wedge \overline{e} = \overline{e} \neq 0$. Since $a \wedge e$ is

nilpotent, there is a smallest integer $n \geq 2$ with $\overline{(a \wedge e)^n} = 0$. But $0 = \overline{(a \wedge e)^n} = \overline{a \wedge e}(a \wedge e)^{n-1} = \bar{e}(a \wedge e)^{n-1} = \overline{e(a \wedge e)^{n-1}} \geq \overline{(a \wedge e)^{n-1}} \geq 0$ contradicts the minimality of n. Thus $a \wedge 2e \leq e$, and so $(a - e)^+ \wedge e = 0$; whence $a \leq e$.

Lemma 3. Let e be a right f-element in R, and suppose that $(a \wedge e)^2 = 0$ implies $a \leq e$ for each positive element a of R. Assume also that S satisfies $x^+x^- = 0$. Then $a \wedge b = 0$ implies that $ab = 0$, where $a \in R$ and $b \in S$.

Proof. Let $a_1 = a - a \wedge e$ and $e_1 = e - a \wedge e$. Since $(a \wedge e) \wedge b = 0$ in S and S satisfies $x^+x^- = 0$, $(a \wedge e)b = 0$. Since $a = a_1 + a \wedge e$, to show that $ab = 0$ it suffices to show that $a_1b = 0$. Since $a_1 \wedge e_1 = 0$ and R_S is an f-module $a_1e_1 \wedge e_1 = 0 = (a_1e_1 \wedge e) \wedge e_1$. But e_1, $a_1e_1 \wedge e \in S$ and hence $e_1(a_1e_1 \wedge e) = 0$. So $0 \leq (a_1e_1 \wedge e)^2 \leq a_1e_1(a_1e_1 \wedge e) = 0$ and hence $a_1e_1 \leq e$. Now $e_1b = (e - a \wedge e)b = eb$, since $(a \wedge e) \wedge b = 0$ in S; and $a_1e_1b = 0$ since $0 \leq a_1e_1 \wedge b \leq ae \wedge b = 0$ and a_1e_1, $b \in S$. But $0 \leq a_1b \leq a_1(eb) = a_1e_1b = 0$ since e is a left superunit of S.

The left annihilator of $X \subseteq R$ is $\ell(X) = \{r \in R: rX = 0\}$. The right annihilator will be denoted by $r(X)$.

Lemma 4. Let e be a left superunit in R and suppose that R has squares positive.

(a) If $x^{2^n} \in S$ for some integer n, then $x^2 \in S$; and if, in addition, x is positive, then $x \in S$.

(b) If ${}_SR_S$ is an f-bimodule, then $x^{2^n} \in S$ implies $x \in S$.

(c) If e is a left identity of R and R_S is a d-module, then $x^{2^n} \in S$ implies $x \in S$.

(d) If e is a superunit of R, then $f^2 \leq f$ implies $f \in S$.

(e) If $e' = (e^2)'$, then $e \wedge x = 0$ implies $nx \leq x^2$ for each integer n.

(f) If R_S is a d-module and $e' = (e^2)'$, then e is a weak order unit of R if and only if e^2 is a weak order unit in Re.

Proof. (a) We use induction on n. If $(x^2)^{2^n} = x^{2^{n+1}} \in S$, then $x^4 \in S$ by induction. But $0 \leq (x^2 - e)^2 = x^4 - (ex^2 + x^2e) + e^2$ yields $0 \leq x^2 \leq ex^2 \leq ex^2 + x^2e \leq x^4 + e^2$. So $x^2 \in S$. If $x \in R^+$, then repeating this argument with x^2 replaced by x gives $x \in S$.

(b) Because of (a) we may assume that $x^2 \in S$. Then $0 \leq (x - e)^2$, $(x + e)^2$ yields that $|xe + ex| \leq x^2 + e^2$; so $|xe + ex| \in S$. But $|x| \leq |xe + ex|$, since this inequality is valid in any totally ordered f-bimodule ${}_SR_S$ which is a homomorphic image of ${}_SR_S$; and ${}_SR_S$ is embedded in the product of all such f-bimodules.

(c) We first note that since $(ye)^+ = y^+e$, Re is an ℓ-subring of R and $\ell(R) = \ell(e)$ is an ℓ-ideal of R. Since $\ell(R)$ is nilpotent, $\ell(R) \subseteq S$ by (a). Now Re $\cap$ S is the convex ℓ-subring of Re generated by e, and if $x^2 \in S$, then $(xe)^2 \in \text{Re} \cap S$. Since e is the identity element of Re, (b) yields that $xe \in S$. But then $x = xe + (x - xe) \in S$.

(d) $0 \leq (e - f)^2$ gives $ef + fe \leq e^2 + f^2$. So $0 \leq 2f \leq ef + fe \leq e^2 + f^2 \leq e^2 + f$ yields $f \leq e^2$.

(e) If n is a positive integer, then $nx \leq nex \leq nex + nxe \leq n^2e^2 + x^2$ since $0 \leq (ne - x)^2$. So $nx \leq nx \wedge n^2e^2 + nx \wedge x^2 = nx \wedge x^2$, since $e' = (e^2)'$.

(f) If e^2 is a weak order unit in Re and $x \wedge e = 0$ in R, then $xe \wedge e^2 = 0$ in Re. So $xe = 0$, and $0 \leq x \leq x^2 \leq x(ex) = 0$ by (e).

Now to the proof of Theorem 8. (a) implies (b): If $a \wedge e = 0$, then $0 \leq a \leq ea = 0$. By Lemma 2 (b) implies (c), and (c) trivially implies (d). (d) implies (a): Since S is an f-ring it satisfies

$x^+x^- = 0$, and each nilpotent element of S is $\leq e$. Suppose that $a \wedge b = 0$ in R, and let $a_1 = a - a \wedge e$, $e_1 = e - a \wedge e$. Note that $be_1 = be$ and $a_1e_1 = 0$, by Lemma 3; so $e_1a_1 \leq e$ by (d), since $(e_1a_1)^2 = 0$. Since ${}_SR$ is an f-module and $b \wedge a_1 = 0$, $b \wedge e_1a_1 = 0$. Whence, by Lemma 3, $be_1a_1 = 0$ and
$0 \leq ba = b(a_1 + a \wedge e) = ba_1 \leq b(ea_1) = be_1a_1 = 0$.

It is clear that (a) and (d) imply (e); and (e) implies (d) by Lemma 4 (a); also, (e) clearly implies (f). Finally, if $e = 1$, then (f) implies (d) by Theorem 7.

4. Remarks and Examples

We note that Theorem 8 may be viewed as a lifting of the identity $x^+x^- = 0$ from S to R and we ask: If R is an ℓ-ring with a weak order unit e and if the convex ℓ-subring of R generated by e satisfies $x^+x^- = 0$ does R satisfy this identity?

Corollary 2. Let R be a right f-ring with a left f-element e which satisfies $ex \geq x$ for each $x \in R^+$. Then R satisfies the identity $x^+x^- = 0$.

Proof. If $e \geq 0$, then $\bar{e} < 0$ in some nonzero totally ordered homomorphic image $\bar{R}_R$ of R_R. But then $0 \leq \bar{x} \leq \overline{ex} \leq \overline{ex} \leq 0$ for each $x \in R^+$, so $\bar{R} = 0$. Thus e is a left superunit in R. Also, $e \wedge x = 0$, implies $0 = ex \wedge x \geq x \wedge x = x \geq 0$, so e is a weak order unit in R. Now apply Theorem 8.

Corollary 3. A right f-ring R which contains a left f-superunit and which has no nonzero nilpotent ℓ-ideals is an f-ring.

Proof. By Corollary 2 R satisfies $x^+x^- = 0$, and so R is an f-ring by Corollary 1.

An ℓ-ring is called <u>archimedean</u> if $na \leq b$ for all integers n implies $a = 0$.

<u>Corollary 4</u>. Let R be an archimedean ℓ-ring with a left superunit e, and let S be the convex ℓ-subring of R generated by e. Then R is an f-ring if and only if R has squares positive and ${}_SR_S$ is an f-bimodule.

<u>Proof</u>. Suppose that R has squares positive and ${}_SR_S$ is an f-bimodule. Since S is an archimedean f-ring with a superunit, S contains no nonzero nilpotent elements. By Lemma 4(b) R contains no nonzero nilpotent elements. Since e is a right f-element of R, $e' = (e^2)'$; so e is a weak order unit of R by Lemma 4(e). By Theorem 8, then, R satisfies $x^+x^- = 0$ and again R is an f-ring by Corollary 1.

The following examples show that the hypothesis that e be a left f-element is necessary for these results and also that the ℓ-ring in Corollary 2 need not be an ℓ-ring. The first example is a modification of an example of Birkhoff and Pierce [3, p. 62].

<u>Example 1</u>. A right f-ring with a left superunit (left identity) that is not an f-ring. Let R and P be totally ordered domains with P a left and right f-R-bimodule; i.e. ${}_RP_R$ is a bimodule and ${}_RP$ and P_R are f-modules. Assume that $pp' \geq \alpha p' \vee p\alpha > 0$ for all $0 < p$, $p' \in P$ and $0 < \alpha \in R$. Let Q be a left ℓ-module over P and over R such that $\alpha q > 0$ and $pq > 0$ for each $0 < \alpha \in R$, $0 < p \in P$, and $0 < q \in Q$. Suppose that multiplication in $P \cup R \cup Q$ is associative whenever it is defined. If $n \geq 2$ is an integer let T be the abelian group

$$T = P \oplus R \oplus Q \oplus \dots \oplus Q \quad (n \text{ copies of } Q).$$

Define multiplication in T by

$$(p,\alpha, q_1, \ldots, q_n)(p',\alpha', q_1',\ldots,q_n') =$$
$$(npp' + \alpha p' + p\alpha',\alpha\alpha',p(q_1' +\ldots+ q_n') + \alpha q_1',\ldots,$$
$$p(q_1' +\ldots+ q_n') + \alpha q_n').$$

Define a positive cone in T by

$$T^+ = \{(p,\alpha, q_1,\ldots,q_n): p > 0, \text{ or } p = 0 \text{ and } \alpha > 0$$
$$\text{or } p = \alpha = 0, \text{ and each } q_i \geq 0\}.$$

Then T is a right f-ring which satisfies the identity $x^+x^- = 0$ but which is not an f-ring. If $e \in R$ is a left superunit (left identity) in the partial semigroup $P \cup R \cup Q$, then $(0,e,0,\ldots,0) = e_1$ is a left superunit (left identity) in T. Note that $\ell(e_1) = \ell(T) = Q \oplus \ldots \oplus Q$, so that e_1 is not a right superunit.

If $P = Q$ is the ring of polynomials over R in one indeterminate, ordered lexicographically with the highest term dominating, and with zero constant term, then P, Q, and R satisfy the conditions in the previous paragraph.

The next example comes from an idea of Elliot Weinberg.

<u>Example 2</u>. A right f-ring with a superunit that does not satisfy $x^+x^- = 0$. Let F be a totally ordered field and let $R = F[x,y]$ be the free (noncommutative) algebra over F in two variables, totally ordered in such a way that $x > 1$ and $y > 1$. If T is the ideal of polynomials in R with zero constant term, then $T = xR \oplus yR$ as right ideals. Define a positive cone in T by $T^+ = \{xr + ys: r,s \in R^+\}$

Then $x + y$ is a superunit in T, and T is the required example.

An f-ring can certainly have a superunit without possessing a unit. It cannot have a left superunit and a right unit without having a unit, however. More generally, we note the following.

Let R be an ℓ-ring with squares positive, and let e be a left superunit of R with $r(e) = 0$. If S is the convex ℓ-subring of R generated by e, then the following statements are equivalent:

(a) R has a unit.

(b) S has a right unit and $\ell(R)$ is directed (every element is the difference of two positive elements).

(c) S has a right unit and e is a superunit.

Proof. (c) → (a): If f is a right identity element of S, then $e = ef$ implies that $r(f) \subseteq r(e) = 0$. So f is a left identity element of R, and hence S is unital. By Lemma 4(d) S contains all the idempotent elements of R, and hence they are all central elements of R (see the proof of Theorem 7).

(b) → (a): If f is a right identity of S, then, as above, f is a left identity of R. Since $\ell(R)^+ \subseteq S$ by Lemma 4(a), and since $\ell(R)$ is directed, $\ell(f) = \ell(R) = 0$. So f is the identity of R.

The example given after Theorem 3 shows that squares positive is needed for (b) → (a) and (c). The example $R = Qf \oplus Qa$ as an ℓ-group, with multiplication given by $f^2 = f$, $a^2 = a$, $fa = f$, $af = a$, and $e = a + f$ shows that squares positive and $r(e) = 0$ is needed for (c) → (a).

5. Semiperfect ℓ-Rings

A ring R with Jacobson radical N is called semiperfect (local) if R/N is left artinian (a division ring) and idempotents may be lifted from R/N to R. R is called π-regular (left π-regular) if for each $a \in R$ there exists a positive integer n and an $x \in R$ such that $a^n x a^n = a^n$ ($a^n = xa^{n+1}$; equivalently, the chain $Ra \supseteq Ra^2 \supseteq \ldots$ is finite). An algebra over a field is <u>locally finite</u> provided each finitely generated subalgebra is finite dimensional. Note that an algebraic algebra is π-regular [1]. Theorem 9 below is given in [13] for the case where R has a unit.

<u>Lemma 5</u>. Let e be a left identity element and a d-element in the ℓ-ring R. Suppose that each element of the ℓ-subring Re is exceeded by a sum of invertible elements.

(a) R is a d-ring if and only if Re is a d-subring and R is a right d-module over ℓ(R).

(b) R is a d-ring which satisfies $x^+x^- = 0$ if and only if R has squares positive and Re is a d-subring.

(c) If R is an f-ring, then Re is an f-subring and each positive invertible element of Re exceeds ℓ(R). Conversely, if Re is an f-subring, e is an f-element, and each positive nonzero element of Re exceeds ℓ(R), then R is an f-ring.

<u>Proof</u>. (a) Suppose that Re is a d subring of R and R is a right d-module over ℓ(R). Let $b \in R^+$ and write $b = b_1 + b_2$ where $b_1 = be$ and $b_2 \in \ell(e) = \ell(R)$. Since ℓ(e) is an ℓ-ideal of R, $|b_2| \subset \ell(R)$; and, since $0 \leq b \leq b_1 + |b_2|$, b is a d-element of R if and only if b_1 is a d-element of R. Now $0 \leq b \leq a_1 + \ldots + a_n \leq |a_1| + \ldots + |a_n|$ where each a_i is an invertible element of Re. But an element in a d-ring is invertible if and only if its absolute value is invertible. Thus, to show that b is a d-element of R we may assume that $b = b_1$ is a positive invertible element of Re.

Now suppose that $x \wedge y = 0$ in R and c is the inverse of b in Re. Since Re is an f-ring and b is positive, so is c. So

$$0 = x \wedge y = ex \wedge ey = cbx \wedge cby \geq c(bx \wedge by) \geq 0,$$

and hence $c(bx \wedge by) = 0$. Thus

$0 = bc(bx \wedge by) = e(bx \wedge by) = bx \wedge by$, and b is a left d-element of R. Also $xb \wedge yb = xeb \wedge yeb = (xe \wedge ye)b = (x \wedge y)eb = 0$, since Re is a d-ring and e is a right d-element of R. So b is a right d-element.

The converse is trivial.

(b) If R has squares positive, then, by Lemma 4(a) or (c), each positive nilpotent element of R is a d-element; so (b) follows from (a) and Theorem 3(d).

(c) If R is an f-ring, then certainly Re is an f-subring. Let a be a positive invertible element of Re and let $b \in \ell(R)^+$. If $\bar{a} \leq \bar{b}$ in some nonzero totally ordered homomorphic image $\bar{R} = R/T$ of R, then $a \leq b + t$ for some $t \in T^+$. If c is the inverse of a in Re, then $0 < e = ac \leq tc \in T$ implies that $\bar{R} = 0$. So $\bar{a} > \bar{b}$ and hence $a > b$.

Conversely, since e is a d-element and $e > \ell(R)$, R is a d-ring by (a), and, since $r(R) = 0$, R is a left f-ring. Suppose that $x \wedge y = 0$ in R and $b \in R^+$. As in the proof of (a), to show that b is a right f-element, we may assume that $b \in Re$. And since $xb \wedge ye = xeb \wedge ye = 0$, we only have to verify that $xb \wedge |y - ye| = 0$. But $xe \wedge |y - ye| = 0$ yields that $xe = 0$ or $xe > |y - ye| = 0$. In either case $xb \wedge |y - ye| = xeb \wedge |y - ye| = 0$.

<u>Theorem 9</u>. Let R be an ℓ-ring with squares positive and suppose the left identity element e of R is an f-element.

(a) If R is semiperfect, π-regular, or left or right π-regular, then R is a d-ring which satisfies $x^+x^- = 0$.

(b) If R is an algebraic ℓ-algebra over a partially order field, then R is locally finite.

(c) If, in addition to the condition in (a), R has the property that $x \in R^+\backslash\ell(R)$ implies that $x > \ell(R)$, then R is an f-ring.

Proof. (a) Since Re is an ℓ-subring of R which inherits the properties in (a), Re is an f-ring by [13]. If Re is local and $a \in R^+e$, then either a or a - e is invertible. Since a = e + (a - e), in either case, a is a sum of units. If R is semiperfect, then Re is a direct sum of local ℓ-rings, so each positive element of Re is exceeded by a sum of units. Whence the conclusion follows from Lemma 5.

If R is π-regular and $a \in R^+e$, let $b = a \vee e$. Then $b^n = b^n x b^n$ implies that $b^n(e - xb^n) = 0$. But $b^n \geq b \geq e$, so b^n has no annihilator in Re. Hence $e = xb^n$ and b is a unit of Re. Again, Lemma 5 applies. A similar argument can be used if R is left or right π-regular.

(b) $Re \cong R/\ell(R)$ is locally finite by [13] and $\ell(R)$ is locally finite since it is nil; so R is locally finite [8; p. 24].

Finally (c) follows from Lemma 5.

References

[1] R. F. Arens and I. Kaplansky, "Topological representation of algebras," Trans. Amer. Math. Soc. 63 (1948), 457-481.

[2] G. Birkhoff, "Lattice Theory," (3rd ed.) Am. Math. Soc. Providence, 1967.

[3] ________ and R. S. Pierce, "Lattice-ordered rings," An. Acad. Brasil. Ci. 28 (1956), 41-69.

[4] J. E. Diem, "A radical for lattice-ordered rings," Pac. J. Math. 25 (1968), 71-82.

[5] N. Divinsky, "Rings and "Radicals," University of Toronto Press, Toronto, 1965.

[6] L. Fuchs, "Teilweise geordnete algebraische Strukturen," Vandenhoeck and Ruprecht, Gottingen, 1966.

[7] M. Henriksen and J. Isbell, "Lattice-ordered rings and function rings," Pac. J. Math. 12 (1962), 535-565.

[8] N. Jacobson, "Structure of Rings," rev. ed., Colloquium Publications No. 36, Amer. Math. Soc., Providence, 1964.

[9] D. G. Johnson, "A structure theory for a class of lattice-ordered rings," Acta. Math. 104 (1960), 163-215.

[10] M. A. Shatalova, "ℓ_A and ℓ_I rings," Siber. Math. J. 7 (1966), 1084-1094.

[11] ________, "The theory of radicals in lattice-ordered rings," Math. Notes of the Acad. of Sciences of the U.S.S.R. 4 (1968), 875-880.

[12] S. A. Steinberg, "Finitely-valued f-modules," Pac. J. Math. 40 (1972), 723-737.

[13] ________, "On lattice-ordered rings in which the square of every element is positive," to appear in J. Australian Math. Soc.

[14] ________, "Radical theory in lattice-ordered rings," Proceedings of the conference on ordered groups and rings, Rome, 1975, to appear.

SEMIGROUP METHODS IN RING THEORY

M. Satyanarayana

Bowling Green State University
Bowling Green, Ohio

Let $(R,+,\cdot)$ be an associate ring. Then $(R,\cdot)$ is a semigroup denoted by $M(R)$. The natural questions one may ask are:

i) Does the semigroup structure of $M(R)$ characterize R? What semigroup theoretic properties influence the structure of a ring?

ii) Given semigroup $(S,\cdot)$, when can it admit ring structure? This consideration incidentally gives the necessary relationship between the additive and multiplicative structure of a ring. Can the addition on $(S,\cdot)$ be unique?

For many results in ring theory, the addition appears to be extraneous, especially with regard to the manipulation of the results involving the properties of intersection of ideals and annihilator properties, etc. The theories concerning these problems are still in an infant stage. The present talk is devoted to the recent advances in obtaining the solutions of the above problems. Among others, Gluskin, Petrich considered the first problem. In some of my papers, I dealt with the above two problems. About unique addition one may refer to Johnson, Ricart, Stephenson and Martindale.

<u>Rees Matrix Semigroup</u>. Let I and Λ be non-empty sets, G be a group, 0 an element not in G and let $G^\circ = G \cup 0$ with 0 acting as zero.

Let $P:\Lambda \times I \to G^0$ with $p_{\lambda i}$ the value at (λ,i), be any function such that for every $i \in I$ there exists $\lambda \in \Lambda$ such that $p_{\lambda i} \neq 0$ and for every $\lambda \in \Lambda$ there exists $i \in I$ such that $p_{\lambda i} \neq 0$. $M^\circ(I,G,\Lambda;P)$ is the set $(I \times G \times \Lambda) \cup 0$ with multiplication $(i,g,\lambda)(j,h,\mu) = (i,gp_{\lambda j}h,\mu)$. This $M^\circ(I,G,\Lambda;P)$ is called Rees matrix semigroup.

Definition. A semigroup is said to be o-simple if it has no ideals except 0 and itself.

Definition. A semigroup is called completely o-simple if it is isomorphic to a Rees matrix semigroup or equivalently it is o-simple semigroup with every idempotent being primitive.

Extensions. A semigroup K is an extension of a semigroup S if S is an ideal of K. A congruence σ on a semigroup S is an equivalence relation σ on S when the relation $a \sigma b$ implies $ac \sigma bc$ and $ca \sigma ca$ for all $c \sigma S$. An extension K of S is dense if the equality relation on K is the only congruence on K whose restriction to S is the equality relation on S. A dense extension K of S is maximal if there exists no dense extension K' of S containing K as a proper subsemigroup. A ring K is an extension of a ring R if R is a subring of K. K is called an essential extension of R if for every ideal J of K, $J \cap R \neq 0$.

Definition. The socle of a ring is the union of its minimal right ideals. The socle of a semigroup is defined similarly.

In a semiprime ring R, minimal one-sided ideals of R coincide with those of M(R), and the socle of R is a two-sided ideal and the socle of R is the socle of M(R).

Definition. A ring R is atomic if R coincides with its socle.

Theorem 1. The following conditions on a ring R are equivalent:

i) R is a semiprime ring with a minimal ring ideal A such that the right annihilator of A is 0;

ii) R is a primitive ring with a non-zero socle;

iii) R is a prime ring with a non-zero socle;

iv) R is an essential extension of its simple socle;

v) R is isomorphic to a dense ring of linear transformations on a vector space containing a non-zero transformation of finite rank;

vi) M(R) is a dense extension of a completely o-simple semigroup;

vii) M(R) has a completely o-simple ideal contained in every non-zero ideal of M(R).

Proof. Combine the results of Gluskin [3], Jacobson [6] and Petrich [9].

Definition. For a division ring Δ and non-empty sets I and Λ any function $P = (p_{\lambda i})$ mapping $\Lambda \times I$ into Δ is a $\Lambda \times I$-matrix over Δ. P is a left row-independent if for any finite $\lambda_1, \lambda_2, \ldots, \lambda_n \in \Lambda$ and scalars $\sigma_1, \sigma_2, \ldots, \sigma_n \in \Delta$, the relations $\Sigma \sigma_j p_{\lambda_j i} = 0$ for every $i \in I$ implies $\sigma_1 = \sigma_2 = \ldots = \sigma_n = 0$. Right-column independence is defined analogously.

Rees-Matrix Ring. Let I, Λ be non-empty sets, Δ be a division ring, $P = (p_{\lambda i})$ be left row-independent and right column-independent $\Lambda \times I$-matrix over Δ; by $M(I,\Delta,\Lambda;P)$ denote the set of all $I \times \Lambda$-matrices over Δ having only a finite number of non-zero entries, together with addition by entries $(a_{i\lambda}) + (b_{i\lambda}) = (a_{i\lambda} + b_{i\lambda})$ and multiplication $*$ given by $A * B = APB$, where the multiplication on the right is the usual row-by-column multiplication of matrices. The ring thus obtained is called Rees-matrix ring.

We need some preliminaries for stating the next theorem. (V,Δ) denotes a right vector space over a division ring . Similarly, (Δ,V) denotes left vector space. Let (,) be a bilinear form on (V,Δ) and let U be a subspace of V. b is called an adjoint of a in U or a is an adjoint of b in V if $(v,bu) = (va,u)$, $u \in U$ and $v \in V$. V^* is the set of all linear transformations on (Δ,V). A subspace U^* of V^* is a t-subspace if for every $0 \neq x \in V$, there exists $f \in U^*$ such that $xf \neq 0$. $L(V)$ is the ring of all endomorphisms on (Δ,V); $F_U(V) = F_U(\Delta,V)$ $= \{a \in L(V)$: a has adjoint in U and rank a is finite$\}$.
$L_U(V) = L_U(\Delta,V) = \{a \in L(V)$: a has adjoint in $U\}$.

<u>Theorem 2</u>. The following conditions on a ring R are equivalent:

i) R is a simple ring with a minimal right ideal;

ii) R is a prime atomic ring;

iii) R is isomorphic to a dense ring of linear transformations of finite rank;

iv) $M(R)$ has a completely o-simple ideal T such that the ring ideal generated by T is R;

v) R is isomorphic to a Rees matrix ring.

<u>Proof</u>. This is proved by showing that T is contained in every non-zero ideal of $M(R)$ and thus R is isomorphic to a dense ring of linear transformations on a vector space containing a non-zero transformation of finite rank. Then the equivalence of all the above conditions can be shown by combining Theorem 1 and the fact $F_U(V) \subseteq R \subseteq L_U(V)$ for some t-subspace U of V^*.

Definition. A semigroup S with 0 is an orthogonal sum of its subsemigroups S_α if $S = \cup S_\alpha$ and $S_\alpha S_\beta = S_\alpha \cap S_\beta = 0$ if $\alpha \neq \beta$.

Definition. A semigroup which is an orthonal sum of completely o-simple semigroups is called a primitive regular semigroup.

Theorem 3. The following conditions on a ring R are equivalent:

i) R is a semiprime atomic ring;

ii) R is a semiprime ring satisfying d.c.c. for principal right ideals;

iii) R is a direct sum of simple atomic rings;

iv) $M(R)$ has a primitive regular ideal S such that R is the ring ideal generated by S.

Proof. This is proved by combining Theorem 2 and the results of Faith [2].

Theorem 4. The following conditions on a ring R are equivalent:

i) R is a semiprime ring which is an essential extension of its non-zero socle;

ii) Every non-zero ideal of R contains an idempotent minimal right ideal;

iii) $M(R)$ is a dense extension of a primitive regular semigroup;

iv) R is isomorphic to a subring of πR_α, where each R_α is a prime ring containing minimal right ideals;

v) $M(R)$ has a primitive regular ideal having non-zero intersection with every non-zero ideal of $M(R)$.

Proof. Let G be the socle of $M(R)$. Set $G = \bigcup_{i \in I} R_i$. On I define a relation θ by: $i \equiv j$ iff $R_i \approx R_j$ as R-modules. Then θ is an equivalence relation. Let $J = I/\theta$ and $R_\alpha = \cup R_i$ $(\alpha \in J)$. Then each R_α is completely o-simple ideal of M(R) and each $G_\alpha = \Sigma R_i$ $(\alpha \in J)$ is a simple atomic ideal. Using this fact and the results of Petrich on bitranslations, the proof is completed.

Theorem 5. The following conditions on a ring R are equivalent:

i) R is a semiprime ring which is a maximal essential extension of its non-zero socle;

ii) R is a complete direct sum of maximal primitive rings;

iii) M(R) is a maximal dense extension of a primitive regular semigroup;

iv) Every non-zero ideal contains an idempotent minimal right ideal.

The proof of this theorem depends upon a result due to Grillet and Petrich [4] which states that if K is an extension of a weakly reductive semigroup S and if $\tau = \tau(K{:}S)$, then K is a dense extension of S iff τ is one-to-one, and is a maximal dense extension iff τ also maps K onto $\Omega(S)$. If we identify S with $\pi(S)$, then dense extensions of S may be identified with all subsemigroups of $\Omega(S)$ containing $\pi(S)$, and maximal dense extensions with $\Omega(S)$, where $\Omega(S)$ is translation hull of S and $\pi(S)$ is the set of all inner bitranslations.

All the above theory is made possible because of the consideration of non-zero socle. If the socle is zero, different methods have to be adopted. In the following I shall indicate some results which are derived from different techniques. All the following results are found in [11] and [12].

Theorem 6. The following are equivalent on a ring R:

i) M(R) is regular and is a N-semigroup (a semigroup is called N-semigroup, if for non-zero x and y, $xy = 0$ implies $x^2 = 0$);

ii) M(R) is an o-simple N-semigroup containing a non-zero idempotent;

iii) M(R) is completely o-simple;

iv) R is a division ring;

v) M(R) has no zero divisors and is right Noetherian.

Theorem 7. If M(R) satisfies right d.c.c. condition and if $x \neq xy$ for every non-zero x and y, then R is nilpotent.

Theorem 8. If $M(R) = \bigcup_{i=1}^{2} x_i \cup x_i M(R)$ and $x_i \notin x_j R$ for $i \neq j$, then R is a finite commutative nilpotent ring.

Problem. If the number of generators is 3, find the structure of the ring.

Theorem 9. Let R be a ring with identity. Let T be the unique maximal ideal of M(R) such that $T = \bigcup_{i=1}^{n} x_i M(R)$. Suppose $x_i x_j = 0$ and $x_i \notin x_j R$ for $i \neq j$. Then the following are true:

i) If R is commutative and $n = 2$, R is a direct sum of fields;

ii) Let R be commutative without idempotents except 0 and 1 or $\bigcap_{i}^{\infty} x_i^t R = 0$ for $i = 1,2,\ldots,n$. If $n \geq 3$ then R is a local ring with every non-unit nilpotent element of index 2. The case $n = 2$ is inadmissible.

iii) If M(R) is a N-semigroup, $n = 2$ is inadmissible.

Theorem 10. Suppose M(R) contains a unique maximal right ideal T such that every right ideal of M(R) is contained in T and R does not contain an identity. Then the following are true:

i) If $R = R^2$, $|R| \leq 2$.

ii) If $|R| > 2$ and R is commutative, T cannot be a ring ideal.

iii) If R is commutative and $T = x \cup xM(R)$, then $|R| \leq 2$.

Recent methods of Weissglass [13] treating the direct sum decomposition of a ring as a special case of semilattice supplementary sum, indicate that there are many possibilities of semigroup methods in ring theory. In fact, Weissglass obtained some properties of semigroup rings through the concept of semilattice supplementary sum.

Finally we conclude with the note that Ecstein used semigroup methods while studying the lifting of orthogonal set of idempotents in topological rings.

References

[1] Ecstein, F., "Semigroup methods in ring theory," J. Algebra 12 (1969), 177-190.

[2] Faith, C., "Rings with minimum condition on principal ideal," Archiv. Math., 10 (1959), 327-330.

[3] Gluskin, L. M., "Semigroups and rings of endomorphisms of linear spaces," Amer. Math. Translations, 45 (1965), 105-137.

[4] Grillet, P. A. and Petrich, M., "Ideal expressions of semigroups," Pacific J. Math., 26 (1968), 493-508.

[5] Jacobson, N., "Structure theory of simple rings without finiteness conditions," Trans. Amer. Math. Soc., 57 (1945), 228-245.

[6] Jacobson, N., "Structure of rings," Amer. Math. Soc. Coll. Pub. Vol. 37, Providence, 1956.

[7] Jaffard, P., "Détermination de certains anneaux," C. R. Acad. Sci., Paris, 229 (1949), 805-806.

[8] Martindale, W. S., "When are multiplicative mappings additive?" Proc. Amer. Math. Soc., 21 (1949), 695-698.

[9] Petrich, M., "Dense extensions of completely o-simple semigroups," J. Reine-Agnew. Math., Part I, 258 (1973), 103-125; Part II, 259 (1973), 109-131.

[10] Ricart, C. E., "One-to-one mappings of rings and lattices," Bull. Amer. Math. Soc., 54 (1948), 758-764.

[11] Satyanarayana, M., "On semigroups admitting ring structure," Semigroup Forum, Part I, 3 (1971), 43-50; Part II, 6 (1973), 189-197.

[12] _______, "On left-cancellative semigroups," Semigroup Forum, 6 (1973), 317-329.

[13] Weissglass, J., "Semigroup rings and semilattice sums of rings," Proc. Amer. Math. Soc., 39 (1973), 471-482.

INJECTIVE HULLS OF SIMPLE MODULES

OVER GROUP RINGS

Robert L. Snider

Virginia Polytechnic Institute and State University
Blacksburg, Virginia

The work of Hall [2] and Roseblade [6] has indicated that integral group rings of polycyclic-by-finite groups share many of the properties of commutative Noetherian rings. With this philosophy in mind, we ask if the injective hull of a simple module over such a group ring is Artin. For commutative Noetherian rings, this was first noted by Matlis [4]. We are able to answer the question affirmatively in the case G is nilpotent-by-finite. Here we are able to follow many of the ideas used in the commutative case. We have no results for general polycyclic-by-finite groups.

We denote the nth term of the socle sequence of a module M by $\mathrm{Soc}_n(M)$ and the injective hull of M by E(M).

Lemma 1. If G is a polycyclic-by-finite group and S a simple Z[G] module, then $E(S) = \bigcup_{n=1}^{\infty} \mathrm{Soc}_n(E(S))$.

Proof. A finitely generated submodule of E(S) is finite [3].

Lemma 2. If G is a finitely generated nilpotent group and M a finitely generated essential extension of a simple Z[G] module S, then any simple factor of M is isomorphic to S.

Proof. S is finite, hence a vector space over Z/pZ for some prime p. Localizing at p, we may assume E(S) and hence M is a module over

$Z_p[G]$. M is finite [3]. Therefore the kernel of the action of G on M has finite index in G and hence we may regard M as a $Z_p[G/H]$ module. Finally $\mathrm{Ann}_{Z_p}(M) = p^n Z_p$ and hence M is a $Z/p^n Z[G/H]$ module. Since G/H is nilpotent, each block of Z/pZ[G/H] contains exactly one simple. This implies each block of R[G/H] contains exactly one simple where R is the p-adic integers [1, p. 300]. We can now pass to $Z/p^n Z[G/H]$ to conclude again that each block contains exactly one simple. This implies each simple factor of M is isomorphic to S.

Theorem. If G is a finitely generated nilpotent group and S a simple Z[G] module, then E(S) is Artin.

We will follow the idea of the commutative proof given in Vamos and Sharpe [7]. We recall that a module is finitely embedded if it has finitely generated essential socle.

Lemma 3. [7, p. 75]. A module is Artin if and only if each homomorphic image is finitely embedded.

Proof of the theorem. Suppose E(S) is not Artin. Pick an ideal I of Z[G] maximal with respect to $\mathrm{Ann}_{E(S)}(I)$ is not Artin. This can be done since Z[G] is Noetherian. The injective hull of S as an $R = Z[G]/I$ is $\mathrm{Ann}_{E(S)}(I)$. We work now over the ring R. We assume E(S) is not Artin but $\mathrm{Ann}_{E(S)}(J)$ is Artin for every nonzero ideal J of R. Since E(S) is not Artin, there exists by Lemma 3 a submodule M with E(S)/M not finitely embedded. Let $\mathcal{A}$ be the set of submodules M with E(S)/M not finitely embedded. We partial order $\mathcal{A}$ by reverse inclusion. Let $\{M_\alpha\}$ be a chain in $\mathcal{A}$ and let $M = \bigcap_\alpha M_\alpha$. If $M \neq M_\alpha$ for any α, then E(S)/M cannot have finitely generated essential socle. If $M = M_\alpha$ for some α, we also have E(S)/M is not finitely embedded. Therefore $\mathcal{A}$ has an upper bound and there is a submodule M of E(S) minimal with respect to E(S)/M is not finitely embedded. M is Artin since if N is a submodule of M, E(S)/N and hence M/N is finitely embedded. Since, by Lemma 1, E(S) is the union of its socle sequence, E(S)/M has essential socle. Since E(S)/M is not finitely embedded, its socle T/M must not be finitely generated. Let $J = \mathrm{Ann}_R(S)$. J contains a nonzero central element c [5]. By Lemma 2, each simple of T/M is isomorphic to S. Therefore $cT \subseteq M$. Multiplication of T by c is an R homomorphism. Therefore we see that T is an extension of $\mathrm{Ann}_T(c)$ by cT. cT and $\mathrm{Ann}_T(c)$ are both Artin and hence T is Artin. Therefore T/M is finitely generated, a contradiction.

Corollary. If G is nilpotent-by-finite and S is simple Z[G] module, then E(S) is Artin.

Proof. Let H be a nilpotent subgroup of finite index. By Clifford's theorem, S is a finite direct sum of simple Z[H] modules. Therefore the injective hull M of S as a Z[H] module is Artin. $\mathrm{Hom}_{Z[H]}(Z[G],M)$ is an injective Z[G] module containing $S = \mathrm{Hom}_{R[G]}(R[G],S)$. As an Z[H] module $\mathrm{Hom}_{Z[H]}(Z[G],M) \simeq \sum_{i=1}^{n} \mathrm{Hom}_{Z[H]}(Z[H]g_i,M)$ where $\{g_1,\ldots,g_n\}$ is a transversal of H in G. $\mathrm{Hom}_{Z[H]}(Z[H]g_i,M) \simeq g_i^{-1} \otimes M$ which is an Artin Z[H] module. Therefore $\mathrm{Hom}_{Z[H]}(Z[G],M)$ is Artin as an Z[H] module and hence as a Z[G] module.

References

[1] L. Dornhoff, "Group Representation Theory," Part B, Marcel Dekker, New York, 1972.

[2] P. Hall, "On the finiteness of certain soluble groups," Proc. London Math. Soc. (3) 9 (1959), 595-622.

[3] A. V. Jategaonkar, "Integral group rings of polycyclic-by-finite groups," J. Pure Appl. Algebra 4 (1974), 337-343.

[4] E. Matlis, "Injective modules over Noetherian rings," Pacific J. Math. 8 (1959), 511-528.

[5] J. Roseblade, "The integral group rings of hypercentral groups," Bull. London Math. Soc. 3 (1971), 351-355.

[6] ________, "Group rings of polycyclic groups," J. Pure Appl. Algebra 3 (1973), 307-328.

[7] D. Sharpe and P. Vamos, "Injective Modules," Cambridge University Press, 1972.

THE ADDITIVE STRUCTURE OF A RING AND THE SPLITTING OF THE TORSION IDEAL

Christine Ayoub

Pennsylvania State University
University Park, Pennsylvania

(Dedicated to the memory of my father, W.L.G. Williams, who died January 31st, 1976)

If the maximal torsion ideal T of a ring A is a ring-theoretic direct summand of A, then A is said to be fissile (German "spaltbar"). F. Szasz [6] proved that every Artinian ring is fissile. In two recent papers ([2] and [3]) Dinh Van Huynh has raised the question whether a ring with minimum condition on principal right ideals (a Min-PR ring) also has this property. He has proved that this is the case if certain extra assumptions are made.

If A is a Min-PR ring, the additive group A^+ is the direct sum of a reduced torsion group B and a divisible group D (cf. [5], p. 424). (In case A is Artinian B is bounded, and the number of components of type P^∞ in D is finite.) We study rings A, whose additive group have this structure, i.e. we assume (*) $A^+ = B \dot{+} D$. (group-theoretic sum), where B is a reduced torsion group and D is divisible. We show that D is an ideal of A and the torsion ideal T of D is a ring-theoretic sum of D, if and only if the torsion ideal T_A of A is a ring-theoretic direct summand of A. Furthermore, we have $A/T_A \approx D/T$. Thus the problem of deciding when a ring with additive structure (*) is fissile reduces to the problem of determining when a divisible ring is fissile. In this case we have the following

Theorem. Let A be a divisible ring and T its torsion ideal. Then T is a ring-theoretic direct summand of A (i.e. A is fissile) if, and only if $T \cap A^2 = 0$.

On the other hand, the following result shows that a mixed divisible ring is not necessarily fissile.

Theorem. Let G be an additive abelian group which is neither a torsion group nor a torsion-free group. Then there exists a ring A on G which is not fissile. (i.e. A is a ring with $A^+ \approx G$).

In the positive direction we have

Theorem. Let A be a divisible ring with torsion ideal T and let $\overline{A} = A/T$. Then $\overline{A}$ is torsion-free and divisible; if for $\overline{a} \in \overline{A}$, $\overline{a} \in \overline{aA}$, then A has the unique ring-theoretic decomposition $A = T \boxplus A^2$ (ring-theoretic direct sum).

If the ring R satisfies the minimum (maximum) condition on principal right ideals we say that R is a Min-PR (Max-PR) ring.

Theorem. If the ring R is torsion-free and divisible and is either a Min-PR or Max-PR ring, then for $r \in R$, $r \in rR$

Summarizing these results we get

Theorem. Let A be a ring with $A^+ = B \dotplus D$, where B is a reduced torsion group and D is divisible. Let T be the torsion ideal of A. Then if for $\overline{a} \in \overline{A} = A/T$, $\overline{a} \in \overline{aA}$, A has the unique ring-theoretic decomposition $A = T \boxplus D^2$.

Corollary. If A is a Min-PR ring then $A = T \boxplus D^2$, where T is the torsion ideal of A and D is the maximal divisible ideal (or maximal divisible subgroup of A^+); D^2 is the only ideal which is a complement for T.

References

[1] Christine Ayoub, "Conditions for a ring to be fissile," Acta. Math. Sci. Hung., to appear.

[2] Dinh van Huynh, "Über die Frage der Spaltberkeit von MHR-Ringen," Bull. Acad. Polon Sci. 23, #2 (1975), pp. 135-138.

[3] ________, "Über Ringe mit Minimalbedingung für Hauptrechtsideale," Acta Math. Acad. Sci. Hung 26 (3-4) (1975), pp. 245-250.

[4] A. Kertész, "Zur Frage der Spaltbarkeit von Ringen," Bull. Acad. Polon Sci. 12, #2 (1964), pp. 91-93.

[5] F. Szász, "Über Ringe mit Minimalbedingung für Hauptrechtsideale II," Acta Math. Acad. Sci. Hung. 12 (1961), pp. 417-439.

[6] ________, "Über Artinsche Ringe," Bull. Acad. Polon Sci. 11 (1963), pp. 351-354.

ON THE COMMUTATIVITY OF CERTAIN TOPOLOGICAL RINGS[1]

Robert L. Blair

Ohio University
Athens, Ohio

1. Introduction

According to a celebrated theorem of Jacobson, a (discrete) ring A is commutative if for every $x \in A$ there is an integer $n \geq 2$ such that $x^n = x$ ([7, Theorem 11]; see also [8, p. 217]). This paper deals with generalizations of this result to certain topological rings.

By way of background, we cite a question due to Herstein and certain results of Lucke: As in [12], let us call a topological ring A a J-ring in case $x \in \mathrm{cl}\{x^n : n \geq 2\}$ for every $x \in A$. (Thus Jacobson's theorem asserts that every discrete J-ring is commutative.) In [4, p. 762], Herstein asks whether every metrizable J-ring is commutative. This question remains open (although every normable J-ring is, in fact, discrete, and hence commutative; see 4.1(b) below). In [12], Lucke shows that every locally compact J-ring is commutative, and at the same time generalizes several other commutativity theorems (due to Herstein [5, Theorems 1 and 2], [6, Theorem 6] and to Belluce, Herstein, and Jain [1, Lemma 1]) to the locally compact semi-simple case [12, Theorem 4].

[1]Preparation of this paper was supported in part by Ohio University Research Committee Grant No. 535.

The principal contribution of the present paper is a topological generalization of the following theorem of Herstein [4, Theorem 21] (not dealt with in [12]): A (discrete) ring A is commutative if (and only if) for every $x \in A$ there is an integer $n \geq 2$ such that $x^n - x \in Z$, where Z is the center of A. To formulate our result, let us call a topological ring A an H-ring in case $\text{cl}\{x^n : n \geq 2\}$ meets $x + Z$ for every $x \in A$. (In the discrete case, this conforms with the terminology of [8, Chap. 10].) Then every locally compact H-ring is commutative (see 3.1 below). Our proof of 3.1 relies on the Herstein theorem just cited, a structure theorem of Kaplansky for locally compact rings with central idempotents (2.1), and on a lemma involving locally compact Q-rings (3.6). We note that, in contrast to the results of [12], semi-simplicity is not required for 3.1.

Lucke deduces the result on J-rings cited above from a general structure theorem for locally compact J-rings [12, Theorem 2]. In §4 we give a different proof of this structure theorem (4.4) and deduce from it the structure of locally compact bounded J-rings (4.5) and of compact J-rings (4.6).

All of the results of this paper (including 4.4) were obtained independently by the author before the publication of [12].

2. Preliminaries

Throughout, A will denote a topological ring, i.e. a ring which is also a Hausdorff space in which multiplication and subtraction are continuous. The center of A is denoted by Z, and if X and Y are subsets of A, then we set $X \cdot Y = \{xy : x \in X \text{ and } y \in Y\}$.

For $x, y \in A$, we write (as in [9]) $x \circ y = x + y + xy$, and x is quasi-regular in case $x \circ x' = 0 = x' \circ x$ for some $x' \in A$. Following

Kaplansky [9], A is a Q-ring in case the quasi-regular elements of A form an open set in A.

Let $(A_\alpha)_{\alpha \in I}$ be a family of topological rings, and for each $\alpha \in I$ let B_α be an open subring of A_α. By the local direct sum of the family $(A_\alpha)_{\alpha \in I}$ relative to the family $(B_\alpha)_{\alpha \in I}$ is meant the topological ring A described as follows: Algebraically, A is the subring of the product $\Pi_{\alpha \in I} A_\alpha$ consisting of all $(x_\alpha)_{\alpha \in I}$ such that $x_\alpha \in B_\alpha$ for all but finitely many $\alpha \in I$; the subring $\Pi_{\alpha \in I} B_\alpha$ of A is then assigned the product topology, and its neighborhoods of 0 are declared to be a fundamental system of neighborhoods of 0 in A (see e.g. [10, p. 21]). In the special case in which each A_α is discrete and $B_\alpha = \{0\}$ for each α, the local direct sum of $(A_\alpha)_{\alpha \in I}$ relative to $(B_\alpha)_{\alpha \in I}$ is precisely the (discrete) direct sum $\oplus\Sigma_{\alpha \in I} A_\alpha$ of the family $(A_\alpha)_{\alpha \in I}$.

In 2.1 and 2.2 we quote the two main results from the literature that will be used in the sequel.

Theorem 2.1 (Kaplansky [10, Theorem 2]). A locally compact ring, all of whose idempotents are in the center, is a local direct sum of Q-rings.

The following basic result from the theory of topological semigroups is due, independently, to Peck [14], Numakura [13], and Koch [11]; see also [15, 3.1]. (The author first learned of this result from L. W. Anderson, who pointed out its relevance to the problems considered in this paper.)

Theorem 2.2 (Peck-Numakura-Koch). If S is a compact semigroup and if $x \in S$, then the set of all cluster points of the sequence $(x^n)_{n=1}^{\infty}$ is a group.

3. Commutativity of H-rings

Here we prove the main result:

Theorem 3.1. Every locally compact H-ring is commutative.

We first prove five lemmas, some of which are of independent interest.

An element $x \in A$ will be called topologically nilpotent (resp. topologically centralpotent) in case $x^n \to 0$ (resp. $x^n \to z$ for some $z \in Z$).

Lemma 3.2. Every topologically centralpotent element of an H-ring is in the center.

Proof. Let A be an H-ring and $x \in A$, and suppose that $x^n \to z$ for some $z \in Z$ but that $x \notin Z$. By hypothesis, $\mathrm{cl}\{x^n : n \geq 2\}$ meets $x + Z$, so there is $y \in \mathrm{cl}\{x^n : n \geq 2\} = \{x^n : n \geq 2\} \cup \{z\}$ and $z_1 \in Z$ such that $y = x + z_1$. Hence $y = x^{n(1)}$ for some $n(1) \geq 2$. Then $(x^{n(1)})^n \to z$ and $x^{n(1)} \notin Z$, so a repetition of the preceding argument yields $n(2) \geq 2n(1)$ with $x^{n(2)} = x^{n(1)} + z'$ for some $z' \in Z$, and hence $x^{n(2)} = x + z_2$ for some $z_2 \in Z$. Proceeding inductively, we thus obtain $x^{n(i)} = x + z_i$ with $z_i \in Z$ and $n(i+1) \geq 2n(i)$ $(i = 1,2,\ldots)$. But then $x^{n(i)} \to z$ and $z_i \to z - x$ so that (since Z is closed) $z - x \in Z$ and hence $x \in Z$, a contradiction.

Lemma 3.3. If A is an H-ring, then all topologically nilpotent elements and all idempotent elements of A are in the center.

Proof. It is immediate from 3.2 that all topologically nilpotent elements are in the center. The argument for idempotents is then standard: If e is an idempotent and if $x \in A$, then $(ex - exe)^2 = 0$, and hence ex - exe commutes with e. Then ex = exe, and, similarly, xe = exe.

Lemma 3.4. Every H-ring whose center is open is commutative.

Proof. If A is an H-ring and if $x \in A$, then $x + Z$ meets $\mathrm{cl}\{x^n : n \geq 2\}$. But $x + Z$ is open, and hence $x + Z$ meets $\{x^n : n \geq 2\}$. Thus $x^n - x \in Z$ for some $n \geq 2$, and it follows from Herstein's theorem [4, Theorem 21] that A is commutative.

Lemma 3.5. If A is an H-ring with a neighborhood of 0 consisting of topologically nilpotent (or, more generally, topologically central-potent) elements, then A is commutative.

Proof. By 3.2, Z has nonempty interior. Hence Z is open, so A is commutative by 3.4.

Lemma 3.6. If A is a locally compact Q-ring, then A has a neighborhood of 0 consisting of topologically nilpotent elements.

Proof. By hypothesis, the set U of all quasi-regular elements of A is a neighborhood of 0, so there is a compact neighborhood V of 0 with $V \subset -U$. By compactness of V, there is a neighborhood W of 0 such that $W \subset V$ and $W \cdot V \subset V$ (see [9, Lemma 10]). It suffices to show that each $x \in W$ is topologically nilpotent. Let $x \in W$ and let $S = \mathrm{cl}\{x^n : n \geq 1\}$. Then $S \subset V$, and hence S is a compact multiplicative semigroup. By 2.2, the set G of all cluster points in S of the sequence $(x^n)_{n=1}^{\infty}$ is a (multiplicative) group, and the identity element e of G is clearly an idempotent in A. But $-e \in U$, so $-e + y - ey = 0$ for some $y \in A$, and hence $e = 0$. Thus 0 is the unique cluster point of $(x^n)_{n=1}^{\infty}$ in the compact space S, and we conclude that $x^n \to 0$.

Proof of 3.1. Let A be a locally compact H-ring. By 3.3 and 2.1, A is a local direct sum of a family $(A_\alpha)_{\alpha\varepsilon I}$ of Q-rings. Now each A_α is topologically isomorphic to a closed subring of A, so each A_α is itself a locally compact H-ring. By 3.6 and 3.5, each A_α is commutative, and hence A is commutative.

It is easy to see that there exist non-locally compact H-rings (see e.g. 4.2(a) below), but we do not know if "locally compact" can be omitted in 3.1 (or even in the commutativity assertion of 4.1(c)). However, certain other hypotheses on an H-ring do ensure commutativity, for example as in 3.5 or in 3.7 below.

A topological ring A is <u>normable</u> if its topology is defined by a norm, i.e. a function $\|\ \|$ from A into the nonnegative real numbers such that $\|x\| = 0$ if and only if $x = 0$, $\|-x\| = \|x\|$, $\|x+y\| \leq \|x\| + \|y\|$, and $\|xy\| \leq \|x\|\|y\|$ for all $x,y \varepsilon A$. (For a discussion of normability of certain topological rings, see [16].)

<u>Proposition 3.7</u>. Every normable H-ring is commutative.

<u>Proof</u>. Let A be a normable H-ring. The set $\{x \varepsilon A:\|x\| < 1\}$ is a neighborhood of 0 consisting of topologically nilpotent elements, so A is commutative by 3.5.

4. <u>Structure Theorems for J-rings</u>

Recall that A is a J-ring in case $x \varepsilon \mathrm{cl}\{x^n : n \geq 2\}$ for every $x \varepsilon A$. Clearly every J-ring is an H-ring. Some basic properties of J-rings follow quickly from the results of §3:

<u>Proposition 4.1</u>. (a) [12, Lemma 4]. If A is a J-ring, then A has no nonzero topologically nilpotent element.

(b) If A is a J-ring with a neighborhood of 0 consisting of topologically nilpotent elements (e.g. if A is normable or a locally compact Q-ring), then A is discrete.

(c) Every locally compact J-ring is commutative and semi-simple.

Proof. (a) Suppose A is a J-ring with $x^n \to 0$ for some nonzero $x \in A$. Since $x \in cl\{x^n : n \geq 2\}$, $x = x^m$ for some $m \geq 2$. But then $e = x^{m-1}$ is idempotent with $e^n \to 0$. Hence $e = 0$, so $x = 0$, a contradiction.

(b) In view of 3.6 and the proof of 3.7, (b) follows immediately from (a).

(c) Let A be a locally compact J-ring. By 3.1, A is commutative. Moreover, by local compactness, the Jacobson radical R of A is closed [10, Theorem 1], so R is itself a locally compact J-ring. Now every element of R is quasi-regular [8, p. 9], so R is a Q-ring, and hence discrete by (b). But every discrete J-ring is semi-simple (see the proof of [8, Chap. 10, Theorem 1]), and thus $R = \{0\}$.

Remarks 4.2. (a) In connection with 4.1(b), we note that a J-ring that is merely metrizable need not be discrete: Let A be the product of denumerably many nonzero discrete J-rings. Then A is first countable (and hence metrizable [2]) and nondiscrete; and A is a J-ring by 4.3 below. (Cf. Herstein's question in §1.)

(b) Lucke deduces the commutativity of locally compact J-rings (4.1(c) above) from a general structure theorem for this class of rings [12, Theorem 2]. Here we give a different proof of this structure theorem (4.4 below) based essentially on 3.6 and 2.1. We first note, in 4.3, a sufficient condition for a (not necessarily locally compact) ring to be a J-ring. (4.3 is considerably stronger than is actually needed for its application in 4.4.)

Proposition 4.3. If A is a local direct sum of a family $(A_\alpha)_{\alpha\in I}$ of J-rings, and if each A_α has a fundamental system of right ideal neighborhoods of 0, then A is a J-ring.

Proof. Assume that A is the local direct sum of $(A_\alpha)_{\alpha\in I}$ relative to the family $(B_\alpha)_{\alpha\in I}$ of open subrings of the A_α's, and let $x = (x_\alpha)_{\alpha\in I}$ be in A. Then for some finite set $J \subset I$ we have $x_\alpha \in B_\alpha$ for every $\alpha \in I - J$. Let V be a neighborhood of 0 in A, and note that V contains a neighborhood of 0 in A of the form $U = \Pi_{\alpha\in I} U_\alpha$, where, for some finite set K with $J \subset K \subset I$, $U_\alpha = B_\alpha$ for every $\alpha \in I - K$, and, for each $\alpha \in K$, U_α is a right ideal neighborhood of 0 in A_α. It suffices to show that $x^n \in x + U$ for some $n \geq 2$.

Now for each $\alpha \in K$,

(1) $x_\alpha^{n(\alpha)} = x_\alpha + u_\alpha$ for some $n(\alpha) \geq 2$ and some $u_\alpha \in U_\alpha$.

Let $n = \Pi_{\alpha\in K}(n(\alpha) - 1) + 1$, and, for each $\alpha \in K$, $m(\alpha) = \Pi_{\beta\in K-\{\alpha\}}(n(\beta) - 1)$. By (1) and induction on j, it follows easily that

(2) $x_\alpha^n = x_\alpha^{n-jn(\alpha)+j} + \Sigma_{i=0}^{j-1} u_\alpha x_\alpha^{n-(i+1)n(\alpha)+i}$

for all $\alpha \in K$ and all $j = 0,\ldots,m(\alpha)$. Setting $j = m(\alpha)$ in (2), we then have

(3) $x_\alpha^n = x_\alpha + \Sigma_{i=0}^{m(\alpha)-1} u_\alpha x_\alpha^{n-(i+1)n(\alpha)+i}$, where $u_\alpha x_\alpha^0 = u_\alpha$.

By (3), $x_\alpha^n \in x_\alpha + U_\alpha$ for all $\alpha \in K$. But this holds also for all $\alpha \in I - K$, so we conclude that $x^n \in x + U$. The proof is now complete.

Theorem 4.4 (cf. [12, Theorem 2]). If A is a topological ring, then the following are equivalent:

(a) A is a locally compact J-ring.

(b) A is of the form $D \oplus E$, where D is a discrete J-ring and E is a local direct sum of a family of discrete J-rings relative to a family of finite fields.

Proof. (a) $\Rightarrow$ (b): By 3.3 and 2.1, we may assume that A is a local direct sum of a family $(A_\alpha)_{\alpha\in I}$ of Q-rings (relative to a family $(B_\alpha)_{\alpha\in I}$, where each B_α is a subring of A_α). Clearly each A_α is a locally compact J-ring, and hence each A_α is discrete by 4.1(b). Note also that $\Pi_{\alpha\in I}B_\alpha$ is open in the locally compact space A, and hence is itself locally compact; thus there is a finite set $J \subset I$ such that B_α is compact (and hence finite) for all $\alpha \in I - J$. Let $K = \{\alpha \in I: \alpha \in J \text{ or } B_\alpha = \{0\}\}$, and let $L = I - K$. For each $\alpha \in L$, B_α is nonzero, finite, commutative, and semisimple (4.1(c)), so by the Wedderburn-Artin theorem [8, p. 40], we have $B_\alpha = \oplus\Sigma_{\beta\in L_\alpha} F_{\alpha\beta}$, where L_α is a finite set and each $F_{\alpha\beta}$ is a finite field. Let $e_{\alpha\beta}$ be the unit element of $F_{\alpha\beta}$, let $u_\alpha = \Sigma_{\beta\in L_\alpha} e_{\alpha\beta}$, and perform a (right) Peirce decomposition [8, p. 50] to express A_α as the following (internal) discrete direct sum: $A_\alpha = (\oplus\Sigma_{\beta\in L_\alpha} e_{\alpha\beta}A_\alpha) \oplus (1-u_\alpha)A_\alpha$. Let $D_1 = \oplus\Sigma_{\alpha\in K}A_\alpha$, $D_2 = \oplus\Sigma_{\alpha\in L}(1-u_\alpha)A_\alpha$, and $M = \{(\alpha,\beta):\alpha \in L \text{ and } \beta \in L_\alpha\}$. The required rings D and E can now be described as follows: $D = D_1 \oplus D_2$, and E is the local direct sum of the family $(e_{\alpha\beta}A_\alpha)_{(\alpha,\beta)\in M}$ relative to $(F_{\alpha\beta})_{(\alpha,\beta)\in M}$. (A routine verification shows that $\phi:A \to D_1 \oplus D_2 \oplus E$ and $\psi:D_1 \oplus D_2 \oplus E \to A$ are inverse topological isomorphisms, where $\phi(x) = ((x_\alpha)_{\alpha\in K}, ((1-u_\alpha)x_\alpha)_{\alpha\in L}, (e_{\alpha\beta}x_\alpha)_{(\alpha,\beta)\in M})$, and where, for $y = (a,b,c) \in D_1 \oplus D_2 \oplus E$, $(\psi(y))_\alpha = a_\alpha$ (resp. $b_\alpha + \Sigma_{\beta\in L_\alpha} c_{\alpha\beta}$) if $\alpha \in K$ (resp. $\alpha \in L$).)

(b) $\Rightarrow$ (a): The hypotheses of (b) clearly imply that A is locally compact, and A is a J-ring by 4.3.

A topological ring A is <u>bounded</u> (see [9, §4]) if for any neighborhood U of 0 in A there is a neighborhood V of 0 in

A such that $V \cdot A \subset U$ and $A \cdot V \subset U$.

Corollary 4.5. If A is a topological ring, then the following are equivalent:

(a) A is a locally compact bounded J-ring.

(b) A is of the form $D' \oplus E'$, where D' is a discrete J-ring and E' is the product of a family of finite fields.

Proof. (a) $\Rightarrow$ (b): We first represent A in the form $D \oplus E$, where D and E are the rings constructed in the proof of 4.4. Since A is bounded, the local direct sum E is bounded. It follows easily that, in all but finitely many summands of E, the field $F_{\alpha\beta}$ is an ideal of, and hence coincides with, the corresponding ring $e_{\alpha\beta}A_\alpha$. The finite number of exceptions are now moved to the discrete summand of A, and A is left in the form $D' \oplus E'$, where D' is discrete and E' is the product of a family of finite fields.

(b) $\Rightarrow$ (a): Since a finite field is necessarily a J-ring, A is a locally compact J-ring by 4.4. Moreover, the direct sum of a discrete ring and a compact ring is bounded [9, p. 161].

Corollary 4.6. A is a compact J-ring if and only if A is the product of a family of finite fields.

Proof. Assume that A is a compact J-ring. Then A is bounded [9, Lemma 10] so, by 4.5, $A = D' \oplus E'$, where D' is a discrete J-ring and E' is the product of a family of finite fields. But clearly D' is a compact (hence finite) J-ring, and hence (as in the proof of 4.4) either $D' = \{0\}$ or D' is itself the product of (a finite family of) finite fields. The converse is clear from 4.5.

For the structure of locally compact rings that satisfy the Jacobson condition of the first paragraph of §1, see [12, Theorem 3].

References

[1] L. P. Belluce, I. N. Herstein, and S. K. Jain, "Generalized commutative rings," Nagoya Math. J. 27 (1966), 1-5.

[2] G. Birkhoff, "A note on topological groups," Compositio Math. 3 (1936), 427-430.

[3] I. N. Herstein, "A generalization of a theorem of Jacobson," Amer. J. Math. 73 (1951), 756-762.

[4] ________, "A generalization of a theorem of Jacobson III," Amer. J. Math. 75 (1953), 105-111.

[5] ________, "Two remarks on the commutativity of rings," Canad. J. Math. 7 (1955), 411-412.

[6] ________, "A condition for commutativity of rings," Canad. J. Math. 9 (1957), 583-586.

[7] N. Jacobson, "Structure theory for algebraic algebras of bounded degree," Annals of Math. 46 (1945), 695-707.

[8] ________, "Structure of rings," rev. ed., Amer. Math. Soc. Colloquium Publications, Vol. 37, Amer. Math. Soc., Providence, R. I., 1964.

[9] I. Kaplansky, "Topological rings," Amer. J. Math. 69 (1947), 153-183.

[10] ________, "Locally compact rings. II," Amer. J. Math. 73 (1951), 20-24.

[11] R. J. Koch, "On topological semigroups," Tulane University Dissertation, 1953.

[12] J. B. Lucke, "Commutativity in locally compact rings," Pacific J. Math. 32 (1970), 187-196.

[13] S. K. Numakura, "On bicompact semigroups," Math. J. Okayama Univ. 1 (1952), 99-108.

[14] J. E. L. Peck, "An ergodic theorem for a noncommutative semigroup of linear operators," Proc. Amer. Math. Soc. 2 (1951), 414-421.

[15] A. D. Wallace, "The structure of topological semigroups," Bull. Amer. Math. Soc. 61 (1955), 95-112.

[16] S. Warner, "Normability of certain topological rings," Proc. Amer. Math. Soc. 33 (1972), 423-427.

AUTHOR INDEX*

Italic numbers indicate the first page of an article by the person named; other numbers refer to pages where the individual is cited in a list of references of an article in this volume.

*The editor wishes to thank V. K. Goel for compiling this index.